United States Nuclear Regulatory Commission

Protecting People and the Environment

NUREG-0711, Rev. 3

I0482730

Human Factors Engineering Program Review Model

Office of Nuclear Regulatory Research

AVAILABILITY OF REFERENCE MATERIALS
IN NRC PUBLICATIONS

United States Nuclear Regulatory Commission

Protecting People and the Environment

NUREG-0711, Rev. 3

Human Factors Engineering Program Review Model

Manuscript Completed: September 2012
Date Published: November 2012

Prepared by:
J. M. O'Hara*, J. C. Higgins*
S. A. Fleger, P. A. Pieringer

*Brookhaven National Laboratory
Nuclear Science and Technology Department
Upton, New York 11973-5000

S. A. Fleger, NRC Project Manager

NRC Job Code N6765

Office of Nuclear Regulatory Research

ABSTRACT

This document is used by the staff of the Nuclear Regulatory Commission to review the human factors engineering (HFE) programs of applicants for construction permits, operating licenses, standard design certifications, combined operating licenses, and license amendments. The purpose of these reviews is to verify that the applicant's HFE program incorporates HFE practices and guidelines accepted by the staff as described within the twelve elements of an HFE program: HFE Program Management, Operating Experience Review, Functional Requirements Analysis and Function Allocation, Task Analysis, Staffing and Qualifications, Treatment of Important Human Actions, Human-System Interface Design, Procedure Development, Training Program Development, Human Factors Verification and Validation, Design Implementation, and Human Performance Monitoring. Each element encompasses five sections: Background, Objective, Applicant Products and Submittals, Review Criteria, and Bibliography.

CONTENTS

LIST OF FIGURES

LIST OF TABLES

EXECUTIVE SUMMARY

The human factors engineering (HFE) staff of the Nuclear Regulatory Commission (NRC) evaluates the HFE programs of applicants for construction permits (CPs), operating licenses (OLs), standard design certifications (DCs), and combined licenses (COLs). The purpose of these reviews is to verify that the HFE aspects of the plant are developed, designed, and evaluated via a structured analysis founded on HFE principles that are acceptable to the NRC staff. The HFE review covers the HFE design process, the HFE final design, its implementation, and ongoing performance monitoring. Therefore, these reviews support public health and safety by verifying that the plants' designs incorporate HFE practices and guidelines.

The methodology of the NRC's HFE review uses a top-down approach for conducting an NRC safety evaluation, so that the significance of individual topics is seen in relationship to the high-level goal of plant safety. Top-down signifies an approach starting at the highest conceptual levels with the plant's high-level mission goals and works down to details by dividing them into the functions necessary to achieve the goals. Functions are allocated to human and system resources and are separated into tasks. The subsequent analysis of personnel tasks identifies the alarms, displays, controls and task support needs required for performing the task. Tasks are arranged into jobs and assigned to staff positions. Each position is evaluated to verify the workload is acceptable. The alarms, displays, controls and task support needs are design inputs for developing the human system interfaces (HSIs), procedures, and training. The detailed design of the HSI, procedures, and training is the "bottom" of the top-down process. The HFE safety evaluation is broad-based and includes normal and emergency operations, maintenance, tests, inspections, and surveillance work.

The overall purpose of the NRC staff's HFE program review is to verify that:

- The applicant integrates HFE into the development, design, and evaluation of the plant.

- The applicant provides HFE products (e.g., HSIs) that facilitate the safe, efficient, and reliable performance of operations, maintenance, tests, inspections, and surveillance tasks.

- The HFE program and its products reflect state-of-the-art human factors principles [cf. Title 10, Part 50.34(f)(2)(iii), of the *Code of Federal Regulations* (10 CFR 50.34(f)) and 10 CFR 52.47(a)(8)], and satisfy all specific regulatory requirements

The HFE program review model consists of 12 review elements, each of which contains five sections:

- *Background* – Briefly explains the rationale and purpose of each element

- *Objective* – Defines the review objective(s) of the element

- *Applicant Products and Submittals* – Lists the materials to be provided for the NRC's review

- *Review Criteria* – Provides the acceptance criteria for the review elements

- *Bibliography* – Includes a list of documents containing detailed information about the aspect of HFE that the element addresses

Following is an overview of each of the elements.

HFE Program Management

The objective of this element is to verify that the applicant has an HFE design team with the responsibility, authority, placement within the organization, and composition to reasonably assure that the plant design meets the commitment to HFE. Further, a plan should guide the team to ensure that the HFE program is properly developed, executed, overseen, and documented. The HFE program plan describes the HFE elements to ensure that HFE principles are applied to the development, design and evaluation of HSI, procedures, and training.

Operating Experience Review

The main purpose of conducting an operating experience review (OER) is to identify HFE-related safety issues. The OER should provide information on the performance of predecessor designs. For new plants, this may be the earlier designs on which the new one is based. For plant modifications, it may be the design of the systems being changed. The issues and lessons learned from operating experience provide a basis to improve the plant's design; i.e., at the beginning of the design process.

The objective of this element is to verify that the applicant identified and analyzed HFE-related problems and issues in previous designs similar to the current one under review. In this way, the negative features of predecessor designs may be avoided in the current one, while retaining positive features. The OER should consider the predecessor systems upon which the design is based, the technological approaches selected (e.g., if touch-screen interfaces are planned, their associated HFE issues should be reviewed), and the plant's HFE issues.

Functional Requirements Analysis and Function Allocation

The purpose of this element is to verify that the applicant defined those functions that must be carried out to satisfy the plant's safety goals and that the assignment of responsibilities for those functions (function allocation) to personnel and automation in a way that takes advantage of human strengths and avoids human limitations.

The personnel role is examined in two steps: functional requirements analysis, and function allocation (assignment of levels of automation). A functional requirements analysis (FRA) identifies those plant functions that must be performed to satisfy the plant's overall operating and safety objectives and goals: To ensure the health and safety of the public by preventing or mitigating the consequences of postulated accidents. This analysis determines the objectives, performance requirements, and constraints of the design, and sets a framework for understanding the role of controllers (personnel or system) in regulating plant processes.

Function allocation is the assignment of functions to (1) personnel (e.g., manual control), (2) automatic systems, and (3) combinations of both. Exploiting the strengths of personnel and system elements enhances the plant's safety and reliability, including improvements achievable through assigning control to these elements with overlapping and redundant responsibilities. Function allocations should be founded on functional requirements and HFE principles in a structured, well-documented methodology that produce clear roles and responsibilities for personnel.

Task Analysis

The functions allocated to plant personnel define the roles and responsibilities that they then accomplish via human actions (HAs). HAs can be divided into tasks, a group of related activities with a common objective or goal. The objective of this review is to verify that the applicant undertook analyses identifying the specific tasks needed to accomplish personnel functions, and also the alarms, information, control- and task-support required to complete those duties. The results of the task analysis offer important inputs in many HFE activities: (1) The analysis of staffing and qualifications; (2) the design of HSIs, procedures, and training program; and (3) criteria for Task Support Verification (see Human Factors Verification and Validation in Section 11).

Staffing and Qualifications

Plant staffing and staff qualifications are important considerations throughout the design process. Initial staffing levels may be established early in the process based on experience with previous plants, staffing goals (such as for staffing reductions), initial analyses, and NRC regulations. However, their acceptability should be examined periodically as the design of the plant evolves. The objective of reviewing staffing and qualification analyses is to verify that the applicant has systematically analyzed the requirements for the number of personnel and their qualifications that includes gaining a thorough understanding of the task and regulatory requirements.

Treatment of Important Human Actions

Over the past several decades, a goal of the NRC's safety programs has been to use risk analyses to prioritize activities, and to ensure that regulators and licensees alike focus efforts and resources on those activities that best support reasonable assurance of adequate protection of the public's health and safety. HFE programs contribute to this goal by applying a graded approach to plant design, focusing greater attention on HAs most important to safety. Therefore, the objective of this element of an HFE program is to identify those HAs most important to safety for a particular plant design; this is accomplished through a combination of probabilistic and deterministic analyses.

The review's objectives are to verify that the applicant has (1) identified important HAs, and (2) considered human-error mechanisms for important HAs in designing the HFE aspects of the plant. They should minimize the likelihood of personnel error, and help ensure that personnel can detect and recover from any errors that occur.

Human-System Interface Design

The objective of this review element is to evaluate the process used by applicants to translate the functional- and task-requirements to HSI design requirements, and to the detailed design of alarms, displays, controls, and other aspects of the HSI. A structured methodology should guide designers in identifying and selecting candidate HSI approaches, defining the detailed design, and performing HSI tests and evaluations. The review also addresses the formulation and employment of HFE guidelines tailored to the unique aspects of the applicants' design, e.g., a style guide to define the design-specific conventions.

Procedure Development

Procedures are essential to plant safety because they support and guide personnel interactions with plant systems and personnel responses to plant-related events. In the nuclear industry, procedure development is the responsibility of individual utilities. The objective of the NRC procedure review is to confirm that the applicant's procedure development program incorporates HFE principles and criteria, along with all other design requirements, to develop procedures that are technically accurate, comprehensive, explicit, easy to utilize, validated, and in conformance with 10 CFR 50.34(f)(2)(ii). The procedures program is reviewed by NRC staff using SRP Chapter 13.

Training Program Development

Training plant personnel is important in ensuring the safe, reliable operation of nuclear power plants. Training programs aid in offering reasonable assurance that plant personnel have the knowledge, skills, and abilities needed to perform their roles and responsibilities. The objective of the training program review is to verify that the applicant has employed a systems approach for developing personnel training. Training programs are reviewed by NRC staff using SRP Chapter 13.

Human Factors Verification and Validation

Verification and validation (V&V) evaluations comprehensively determine that the final HFE design conforms to accepted design principles, and enables personnel to successfully and safely perform their tasks to achieve operational goals. This element involves three evaluations, with the following objectives:

- *HSI Task Support Verification* - the applicant verified that the HSI provides the alarms, information, controls, and task support defined by tasks analysis needed for personnel to perform their tasks.

- *HFE Design Verification* - the applicant verified that the design of the HSIs conform to HFE guidelines (such as the applicant's style guide).

- *Integrated System Validation* - the applicant validated, using performance-based tests, that the integrated system design (i.e., hardware, software, procedures and personnel elements) supports safe operation of the plant.

These evaluations identify human engineering discrepancies (HEDs). The NRC staff's review of the applicant's HED resolutions verifies that the applicant assessed the importance of HEDs, corrected important ones, and that the corrections are acceptable.

Design Implementation

This element addresses implementation of the HFE aspects of the plant design for new plants and plant modifications. For a new plant, the implementation phase is well defined and carefully monitored through start-up procedures and testing; implementing modifications is more complex.

The objectives of this review are to verify that the applicant's:

- as-built design conforms to the verified and validated design resulting from the HFE design process
- implementation of plant changes considers the effect on personnel performance, and affords necessary support to reasonably assure safe operations

<u>Human Performance Monitoring</u>

The objective of reviewing an applicant's human performance monitoring program is to verify that the applicant prepared a program to:

- adequately assure that the conclusions drawn from the integrated system validation remain valid with time
- ensure that no significant safety degradation occurs because of any changes made in the plant

The applicant may incorporate this monitoring program into their problem identification and resolution program and their training program.

This document is the third revision to NUREG-0711. NUREG-0711 Revision 0 (1994) was published to establish the criteria for reviewing the human factors aspects of design certification submittals for advanced nuclear power plants. Revision 1, published in May 2002, (1) provided additional HFE review guidance for hybrid HSIs; (2) revised the sections on Functional Requirements Analysis and Function Allocation, Human Reliability Analysis, Human-System Interface Design, and HFE Verification and Validation; (3) added new sections on Design Implementation and Human Performance Monitoring; and (4) integrated the NRC's HFE review processes into a single document.

Revision 1 was submitted for public comment in December, 2002. Revision 2 incorporates the changes the NRC made to the document in response to the comments and was published in 2004.

This document, Revision 3, incorporates lessons learned from using NUREG-0711 in several design certification reviews of new plants and includes new guidance published since the last revision, such as NRC staff's interim guidance documents and the results of NRC research projects.

Some of the key technical revisions included in Revision 3 include the following enhancements to the guidelines reviewers use when applying the Standard Review Plan (NUREG-0800) to HFE:

- The "Function Requirements Analysis and Function Allocation" element better addresses modern implementations of automation.
- The name of the former "Human Reliability Analysis" element was changed to "Treatment of Important Human Actions" and its scope was expanded to address human actions that the applicant either identifies deterministically or identifies using risk analysis. Deterministic engineering analyses typically are completed by applicants as part of the suite of analyses in the Final Safety Analysis Report and Design Control Document in Chapter 7, Instrumentation & Control, and Chapter 15, Transients and Accident Analysis. These

deterministic analyses often credit human actions. Also this expansion of scope now incorporates the review of those actions identified in SRP Chapter 18, Appendix 18-A (Guidance for Crediting Manual Operator Actions in Diversity and Defense-in-Depth (D3) Analyses). This revision will ensure more complete and consistent review of these deterministically identified human actions.

- The "Human-System Interface (HSI) Design" element includes new specific guidance for the review of the detailed design and integration of the main control room, technical support center, emergency operations facility, remote shutdown facility, and local control stations. HFE review aspects of the following documents have been included in this element:

  - 10CFR50.34(f)

  - I&C BTP 7-19, Guidance for Evaluation of Diversity and Defense-In-Depth in Digital Computer-Based Instrumentation and Control Systems

  - RG 1.62, Manual Initiation of Protective Actions

  - RG 1.97, Criteria for Accident Monitoring Instrumentation for Nuclear Power Plants

  - NUREG-0654, Criteria for Preparation and Evaluation of Radiological Emergency Response Plans and Preparedness in Support of Nuclear Power Plants

- In the "Human Factors Verification and Validation" element, the guidance on scenario development, performance measurement, and the process by which human engineering discrepancies are evaluated was simplified and consolidated to eliminate redundancy.

- Enhancements to "Task Analysis," "HSI Design," and "HFE Verification and Validation" elements address the controls and displays for manual actions identified in Point 4 of I&C BTP 7-19.

Other changes made in Revision 3 of NUREG-0711 facilitate the use of the guidelines by reviewers, such as adding an "Additional Information" section to some of the review criteria to explain the basis or to give examples to support the reviewers' understanding of the guideline's meaning. In addition, the document was revised in accordance with the NRC's guidance on Plain Language.

ACRONYMS AND ABBREVIATIONS

AIAA	American Institute of Aeronautics and Astronautics
ANS	American Nuclear Society
ANSI	American National Standards Institute
ATWS	anticipated transients without scram
BNL	Brookhaven National Laboratory
BWR	boiling water reactor
CFR	U.S. Code of Federal Regulations
COL	combined license
CP	construction permits
CSF	critical safety functions
D3	diversity and defense in depth
DC	design certifications
DCD	design control document
EOF	emergency operations facility
EOP	emergency operating procedure
FRA	functional requirements analysis
FSAR	final safety analysis report
FV	Fussell-Vesely
GDC	General Design Criteria
GTG	generic technical guidelines
HA	human action
HED	human engineering discrepancy
HFE	human factors engineering
HFES	Human Factors and Ergonomic Society
HRA	human reliability analysis
HSI	human-system interface
I&C	instrumentation and control
IAEA	International Atomic Energy Agency
IEC	International Electrotechnical Commission
IEEE	Institute of Electrical and Electronics Engineers
IP	implementation plan
ISO	International Standards Organization
ITAAC	inspections, tests, analyses, and acceptance criteria
ISV	integrated system validation
LCS	local control station
LOCA	loss-of-coolant accident
MCR	main control room
MUX	multiplexer
NEI	Nuclear Energy Institute
NPP	nuclear power plant
NRC	Nuclear Regulatory Commission
NSSS	nuclear steam supply system
OER	operating experience review
OL	operating license
PRA	probabilistic risk assessment
PSA	probabilistic safety assessment
PSF	performance shaping factor
PWR	pressurized water reactor

RAW	risk achievement worth
RCS	reactor coolant system
RG	regulatory guide
RIS	regulatory issue summary
RSF	remote shutdown facility
RSR	results summary report
SGTR	steam generator tube rupture
SPDS	safety parameter display system
SRP	Standard Review Plan
SSC	structure, system, and component
T&E	test and evaluation
TMI	Three Mile Island
TSC	technical support center
V&V	verification and validation

1   INTRODUCTION

1.1   Background

One important insight from studies of the Three Mile Island (TMI), Chernobyl, and other nuclear power plant (NPP) accidents is that errors resulting from human factors deficiencies, such as poor control room design, procedures, and training are a significant contributing factor to NPP incidents and accidents.

Plant safety requires "defense in depth" that encompasses using multiple barriers to prevent the release of radioactive materials, and employs a variety of programs to assure the integrity of barriers and related systems (IAEA, 1999). These programs include conservative design, quality assurance, administrative controls, and human factors. Human factors engineering (HFE) plays a major role in supporting plant safety and providing defense in depth.

The HFE staff of the Nuclear Regulatory Commission (NRC) evaluates the HFE programs of applicants for construction permits (CPs), operating licenses (OLs), standard design certifications (DCs), combined licenses (COLs), and amendments to licenses. The purpose of these reviews is to support public health and safety by verifying that the applicant's HFE program incorporates HFE practices and guidelines that are acceptable to the NRC staff. The scope of the NRC staff's HFE reviews includes the design process, the final design, its implementation, and ongoing performance monitoring.

General guidance to the NRC staff for the performance of HFE reviews is in Chapter 18 of the Standard Review Plan, NUREG-0800 (NRC, 2007). This document, the *Human Factors Engineering Program Review Model* (NUREG-0711), supports the NRC staff's HFE reviews by detailing the review criteria. The review process reflects a "top-down" approach to conducting an HFE program safety evaluation. "Top-down" denotes that the review approach starts at the "top" with an overview of the high-level plant goals. Then the functions necessary to achieve the plant's goals are defined and sometimes refined in greater detail into systems and subsystems. Functions are allocated to human and system resources and subsequently separated into tasks for specifying the alarms, information, controls, and task support needs needed to complete functional assignments. Tasks are arranged into jobs and assigned to staff positions. Each position is evaluated to verify the workload is acceptable. The alarms, displays, controls and task support needs are design inputs for developing the human-system interface (HSI), procedures, and training. The detailed design of the HSIs is the bottom of the "top-down" process. The HFE safety evaluation is broad-based, covering normal and emergency operations, maintenance, tests, inspections, and surveillance.

NRC regulations in Title 10, Part 50, of the *Code of Federal Regulations* (10 CFR 50) and 10 CFR 52 require personnel to use a variety of controls and displays. They also require a control room that reflects state-of-the-art human factors principles. This document offers detailed guidance for the NRC staff to use in verifying that these requirements are met. NUREG-1649 (NRC, 2000) describes the NRC's Reactor Inspection and Oversight Program for operating reactors. This program is outlined using "cornerstones" for reactor safety, radiation safety, and security; they are initiating events, mitigation systems, barrier integrity, and emergency preparedness. Well-designed HSIs, procedures and training are important to optimizing each of these four cornerstones, and the guidance in this document will help verify that they are well-designed. HSIs, procedures, and training also are important in helping to assure the radiation-safety cornerstone goals of minimizing the radiation exposure of plant

1

workers and the general public during routine operations. These guidelines also are applicable to bettering the functionality of the plant-security program's central- and secondary- alarm stations. In addition, there are three cross-cutting elements in the Reactor Oversight Program, one of which is human performance. One of the principal purposes of this program is assisting the NRC staff verification that HSIs, procedures, and training support human performance.

An applicant's HFE program reasonably assures plant safety when it conforms to the following six principles: (1) Developed by a qualified HFE design team, using an acceptable HFE program plan; (2) derived from suitable HFE studies and analyses that afford accurate and complete inputs to the assessment criteria for the design process, and the verification and validation (V&V) process; (3) designed via proven technology incorporating accepted HFE standards and guidelines; (4) evaluated with a thorough V&V test program; (5) implemented such that it effectively supports operations; and (6) monitored during operations to detect changes in human performance.

1.2   General Description of the Program Review Model

1.2.1   Purpose of an HFE Safety Review

The overall purpose of the NRC's staff's HFE program review is to verify that:

- The applicant integrates HFE into the development, design, and evaluation of the plant.

- The applicant provides HFE products (e.g., HSIs) that facilitate the safe, efficient, and reliable performance of operations, maintenance, tests, inspections, and surveillance tasks.

- The HFE program and its products reflect state-of-the-art human factors principles [cf. 10 CFR 50.34(f)(2)(iii) and 10 CFR 52.47(a)(8)], and satisfy all specific regulatory requirements.

10 CFR 52.47 requires that applications for design certification of new reactor designs meet the technically relevant portions of the TMI requirements in 10 CFR 50.34(f). 10 CFR 50.34(f)(2)(iii) requires that a control room reflects state-of-the-art human factors principles. Also, 50.34 specifically requires several features: A safety parameter display system console; automatic indication of bypassed and operable status of safety systems; and monitoring capability in the control room of a variety of system parameters. 10 CFR 55.46 also necessitates having a plant-referenced simulator capability.

In this document, the state-of-the-art human factors principles are those ones currently accepted by human factors practitioners; here, "current" refers to the time when a plan or product is prepared. "Accepted" is regarded as a practice, method, or guide that is (1) documented in the human factors literature within a standard or guidance document that underwent a peer-review process, or (2) is justified through scientific research and/or industrial practices.

1.2.2 Review Elements

Figure 1-1 illustrates the division of the NRC staff's HFE safety review into 12 elements arranged in four general activities. These elements contain the criteria for reviewing an applicant's submittal describing their HFE program and the resulting design. The NRC staff's review of the applicant's submittal via the NUREG-0711 criteria serves to formulate a safety finding about the acceptability of the applicant's HFE design.

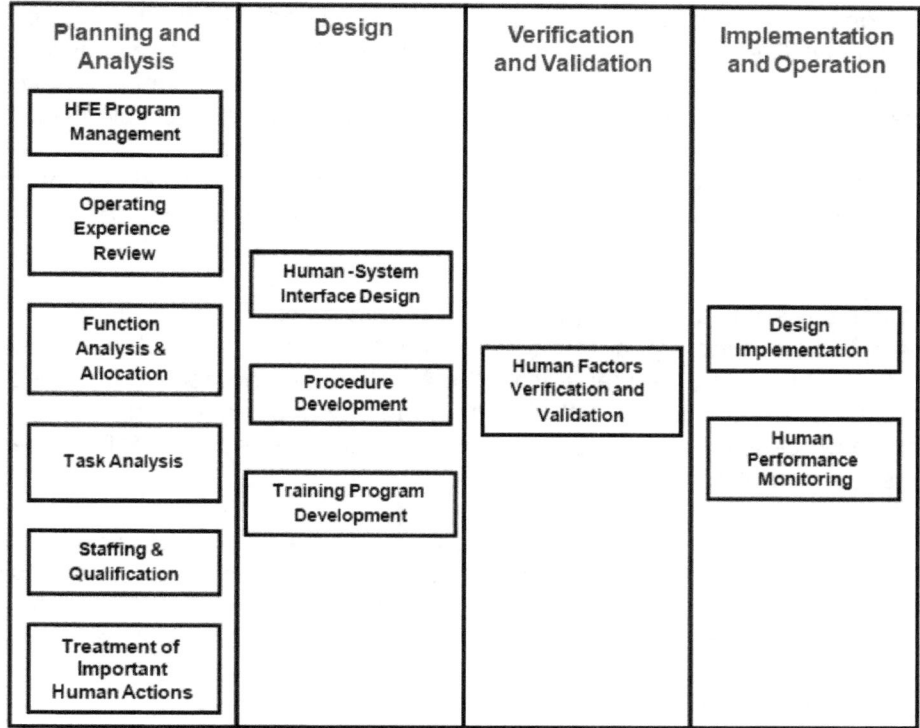

Figure 1-1 Elements of the HFE program's review model

Each element is divided into five sections: Background, Objective, Applicant Products and Submittals, Review Criteria, and Bibliography.

Background

This section explains the rationale underlying, and purpose of the element.

Objective

This section defines the objective(s) of reviewing the element.

Applicant Products and Submittals

Each element lists the types of products that applicants are likely to produce by using their HFE activities. These products may include plans, detailed analysis results, results summary reports, design descriptions, and actual designs, such as the control room HSIs. Some of the applicant's products are submitted to the NRC as part of the licensing review process. These are referred to as submittals. The Final Safety Analysis Report (FSAR) and in the Design Control Document (DCD) are two important submittals. For HFE reviews, implementation plans and results summary reports are two important types of submittals used in the review of HFE elements. As part of the NRC's review process, submittals are evaluated and the staff may review other HFE products to supplement the safety review.

Implementation Plan

An *implementation plan* (IP) describes the applicant's proposed methodology for conducting an HFE element. The NRC staff reviews an IP methodology using the review criteria for the element provided in this document (NUREG-0711). The focus of the staff's review is to reasonably assure that the applicant's methodology will generate acceptable results that satisfy the staff's review criteria. Where the staff's review criteria identify the scope of the HFE analysis, e.g., Task Analysis, Criterion 1, the applicant's commitment to simply meet the criterion is acceptable. However, for other criteria, the applicant should detail the methodology to be employed.

An IP review gives the applicant the opportunity to obtain an NRC staff review of, and concurrence with the methodology before the applicant conducts the work associated with the element. This type of review is desirable from the NRC staff's perspective because it offers the staff an opportunity to identify issues with the methodology and provide input early in the analysis or design process when the applicant more easily can address staff concerns than when the element is completed.

As shown in Figure 1-1, the NRC's HFE review encompasses 12 HFE elements. When the final results for an HFE element are not available for the review (e. g., by design certification), the NRC staff accepts IPs for HFE activities as the basis for making a safety finding for a particular plant design. However, when an applicant uses an IP for design certification, an associated set of "inspections, tests, analyses, and acceptance criteria" is required to ensure completion of the HFE element in accordance with the IP.

Because the IPs are a main basis of the NRC's safety finding for incomplete HFE activities, the NRC staff must understand fully how the applicant's methodology will be implemented, be confident that design personnel will undertake it reliably, and be confident that the results will conform to NUREG-0711 review criteria. To determine whether an IP is acceptable, the NRC staff evaluates whether the IP is:

- complete, i.e., the IP describes the scope, inputs, analyses to be performed, outputs, and documentation

- detailed, i.e., the IP describes the methodology in a step-by-step format to ensure that the applicant's design personnel can reliably use the IP, and that knowledgeable engineers will obtain consistent results from executing the methodology

- verifiable, i.e., the final results can be evaluated using NUREG-0711 criteria, and the IP describes the products (expected results from executing the methodology)

The specific IPs forming the basis of the NRC staff's safety finding for a design certification application should be referenced in the DCD. If the IP is designated Tier 2* (referred to as "Tier 2 star"), any change to the approved IP will require prior NRC approval.

Results Summary Report

A *results summary report* (RSR) summarizes the results of a completed NUREG-0711 element and cites documents or files that contain the complete results. Using the review criteria in this document (NUREG-0711), the NRC staff will evaluate the summary of results for each element.

If an applicant submits an IP prior to the RSR, the RSR should contain sufficient detail for the NRC staff to determine that the results were derived from implementing the methodology contained in the applicant's previously reviewed and approved IP.

If an applicant completed the NUREG-0711 element before the NRC's review of the applicant's methodology and an IP was not developed, then the RSR should describe the methodology used (or refer to a document containing that complete description), as well as a summary of the results derived from implementing that methodology.

Each HFE review element includes the expected content of an RSR for that element.

The applicant may provide IP and RSR documentation in the form of one or more reports. In addition, the NRC staff may audit or inspect the detailed design information at the applicant's facility to supplement the information in the IP and RSR documents.

Review Criteria

This section of each element contains the acceptance criteria for each HFE element. Where appropriate, references are made to more detailed NRC guidance (e.g., NUREG-0700). For some review criteria, "Additional Information" is given, explaining or giving examples to support the reviewers' understanding of the criterion's meaning.

Bibliography

For each HFE element, a list of documents is provided containing detailed information about the aspect of HFE addressed by the element. Full citations to these documents are given in Section 14, References. For some documents that are updated periodically, such as NUREG-0700, the versions referenced may not be the most recent ones.

1.3   Use of This Document

The purpose of the NRC staff's HFE review is to support the NRC's safety mission of protecting people and the environment by verifying that accepted HFE practices and guidelines are incorporated into a plant's physical design and programs. NUREG-0711 presents a review methodology addressing the scope of NRC HFE reviews identified in the Standard Review Plan (SRP), NUREG-0800, Chapter 18, Human Factors Engineering, including the review of:

- HFE aspects of a new plant

- HFE aspects of control room modifications

- HFE aspects of modifications affecting risk-important human actions

In addition, NRC inspectors can use selected elements of the guidance in this document to support their review of those aspects of incidents with important human-performance contributions.

The NRC, the nuclear industry, and the public, have adopted a broader consideration of risk in many activities associated with NPPs. Therefore, the concept of risk importance is integral to the guidance in this document. Applying the precepts of risk importance will help reviewers decide which particular items to review and the depth of those reviews.

The level of NRC staff's review of an applicant's HFE design should also reflect the unique circumstances of the review. For example, a review of a new nuclear power plant will likely use all the elements, while a review of changes to the HSIs of an existing plant will likely use only a subset of the elements. Thus, the NRC staff will tailor the guidance they employ based on such circumstances. NUREG-0800, Chapter 18 explains the grading of HFE reviews.

This document is the third revision to NUREG-0711. NUREG-0711 Revision 0 (1994) was published to establish the criteria for reviewing the human factors aspects of design certification submittals for advanced nuclear power plants. Revision 1, published in May 2002, (1) provided additional HFE review guidance for hybrid HSIs; (2) revised the sections on Functional Requirements Analysis and Function Allocation, Human Reliability Analysis, Human-System Interface Design, and HFE Verification and Validation; (3) added new sections on Design Implementation and Human Performance Monitoring; and (4) integrated the NRC's HFE review processes into a single document.

Revision 1 was submitted for public comment in December, 2002. Revision 2 incorporates the changes the NRC made to the document in response to the comments and was published in 2004.

This document, Revision 3, incorporates lessons learned from using NUREG-0711 in several design certification reviews of new plants and includes new guidance published since the last revision, such as NRC staff's interim guidance documents and the results of NRC research projects.

Some of the key technical revisions included in Revision 3 include the following enhancements to the guidelines reviewers use when applying the Standard Review Plan (NUREG-0800) to HFE:

- The "Function Requirements Analysis and Function Allocation" element better addresses modern implementations of automation.

- The name of the former "Human Reliability Analysis" element was changed to "Treatment of Important Human Actions" and its scope was expanded to address human actions that the applicant either identifies deterministically or identifies using risk analysis. Deterministic engineering analyses typically are completed by applicants as part of the suite of analyses in the Final Safety Analysis Report and Design Control Document in Chapter 7, Instrumentation & Control, and Chapter 15, Transients and Accident Analysis. These deterministic analyses often credit human actions. Also this expansion of scope now incorporates the review of those actions identified in SRP Chapter 18, Appendix 18-A (Guidance for Crediting Manual Operator Actions in Diversity and Defense-in-Depth (D3) Analyses). This revision will ensure more complete and consistent review of these deterministically identified human actions.

- The "Human-System Interface (HSI) Design" element includes new specific guidance for the review of the detailed design and integration of the main control room, technical support center, emergency operations facility, remote shutdown facility, and local control stations. HFE review aspects of the following documents have been included in this element:

- 10CFR50.34(f)

- I&C BTP 7-19, Guidance for Evaluation of Diversity and Defense-In-Depth in Digital Computer-Based Instrumentation and Control Systems

- RG 1.62, Manual Initiation of Protective Actions

- RG 1.97, Criteria for Accident Monitoring Instrumentation for Nuclear Power Plants

- NUREG-0654, Criteria for Preparation and Evaluation of Radiological Emergency Response Plans and Preparedness in Support of Nuclear Power Plants

- In the "Human Factors Verification and Validation" element, the guidance on scenario development, performance measurement, and the process by which human engineering discrepancies are evaluated was simplified and consolidated to eliminate redundancy.

- Enhancements to "Task Analysis," "HSI Design," and "HFE Verification and Validation" elements address the controls and displays for manual actions identified in Point 4 of I&C BTP 7-19.

Other changes made in Revision 3 of NUREG-0711 facilitate the use of the guidelines by reviewers, such as adding an "Additional Information" section to some of the review criteria to explain the basis or to give examples to support the reviewers' understanding of the guideline's meaning. In addition, the document was revised in accordance with the NRC's guidance on Plain Language.

2   HFE PROGRAM MANAGEMENT

2.1   Background

To effectively accomplish HFE in designing and modifying a plant, an applicant should have an HFE program plan that is implemented by a qualified HFE design team. The review criteria in this element address:

- general goals and scope of the HFE program
- HFE team, member qualifications, and organization
- HFE process and procedures
- HFE issues tracking
- HFE elements

2.2   Objective

The NRC staff uses the review criteria in this section to verify that:

- The applicant has an HFE design team with the responsibility, authority, placement within the organization, and qualifications to verify that the plant design commitment to HFE is met.
- The applicant has an HFE program plan that reasonably assures that the HFE is properly developed, executed, overseen, and documented.
- The HFE program plan describes the HFE elements to ensure that HFE principles are applied to the development, design and evaluation of HSIs, procedures, and training.
- The HFE program plan appropriately considers and addresses the deterministic aspects of design, discussed in Regulatory Guide (RG) 1.174.
- The HFE program provides assurance that modifications to the plant do not compromise good human factors design.

2.3   Applicant Products and Submittals

The product of the HFE Program Management element is the HFE Program Management Implementation Plan and a complete description of the HFE program organization that is establishing the HFE elements. The plan should explain the applicant's HFE goals/objectives, technical program to accomplish the objectives, the system to track HFE issues, the qualifications of the HFE design-team members, and the management and organizational structure supporting the accomplishment of the HFE elements. There is no RSR for this element.

2.4   Review Criteria

2.4.1   General HFE Program Goals and Scope

(1)   *HFE Program Goals* – The applicant should state the general objectives of the program in "human-centered" terms. As the HFE program develops, they should be further defined and used as a basis for HFE tests and evaluations.

Additional Information: Generic "human-centered" HFE design goals include the following:

- personnel tasks can be accomplished within time and performance criteria
- the HSIs, procedures, staffing/qualifications, training, and management and organizational arrangements support personnel situation awareness
- the design will support personnel in maintaining vigilance over plant operations and provide acceptable workload levels, i.e., minimize periods of under- and over-load
- the HSIs will minimize personnel error and will support error detection and recovery capability

(2) *Assumptions and Constraints* – The applicant should identify the design assumptions and constraints.

Additional Information: An assumption or constraint is an aspect of the design, such as a specific staffing plan or a specific HSI technology that is an *input* to the HFE program rather than the result of HFE analyses and evaluations.

(3) *HFE Program Duration* – The applicant's HFE program should be in effect at least from the start of the design cycle through completion of initial plant startup test program.

(4) *Facilities* – The applicant's HFE program should cover the main control room (MCR), remote shutdown facility (RSF), technical support center (TSC), emergency operations facility (EOF), and local control stations (LCSs). The 12 HFE elements should be applied to each of them, unless otherwise noted for a specific HFE element. However, applicants may apply the elements of the HFE program in a graded fashion to facilities other than the MCR and RSF, providing justification in the HFE program plan.

(5) *HSIs, Procedures and Training* – The applicant's HFE program should address the design of HSIs and identify inputs to the development of procedures and training for all operations, accident management, maintenance, test, inspections, and surveillance tasks that operational personnel will perform or supervise. In addition, the HFE design process should identify training program input for the following personnel identified in 10 CFR 50.120: instrument and control technician, electrical maintenance personnel, mechanical maintenance personnel, radiological protection technician, chemistry technician, and engineering support personnel. In addition, any other personnel who perform tasks directly related to plant safety should be included, such as information technology technicians who troubleshoot and maintain support systems and their HSIs.

(6) *Personnel* – The applicant's HFE program should consider operations staffing and qualifications, including licensed control-room operators as defined in 10 CFR Part 55, and the following categories of personnel: non-licensed operators, shift supervisor, and shift technical advisor.

(7) *Additional Considerations for Reviewing the HFE Aspects of Plant Modifications* – In addition to any of the criteria above that relate to the modification being reviewed, the applicant should address the following considerations:

- The goals of the applicant's HFE program should address the potential effects of a modification on the performance of personnel. The transition from the existing plant configuration to the modified one can pose different demands on human performance than either the initial or the final configurations. Therefore, the modification and its implementation should be planned to minimize the effects of the change on personnel performance. The HFE program for the modification should consider:

- planning the installation to minimize disruptions to work
- coordinating changes in training and procedures when implementing the modification
- conducting training to maximize personnel's knowledge and skill with the new design before implementing it

- The applicant's HFE program should involve plant personnel to ensure that the following are considered from a user's perspective in establishing the requirements for the modification, and evaluating the outputs of the design process:

- user's understanding of how plant systems are structured and behave
- task demands and constraints of the existing work environment and work processes

2.4.2 HFE Team and Organization

In this document, the term "HFE team" means the primary organization(s) responsible for the applicant's HFE program. However, we do not assume that HFE is the responsibility of a single organizational unit, or that there is an organizational unit called the "HFE team."

(1)   *Responsibility* – The applicant's team should be responsible for:

- developing all HFE plans and procedures
- overseeing and reviewing all activities in HFE design, development, test, and evaluation, including the initiation, recommendation, and provision of solutions through designated channels for problems identified in implementing the HFE work
- verifying that the team's recommendations are implemented
- assuring that all HFE activities comply with the HFE plans and procedures
- scheduling work and milestones

(2)   *Organizational Placement and Authority* – The applicant should describe the primary HFE organization(s) or function(s) within the engineering organization designing the plant or modification. The organization should be illustrated to show organizational and functional relationships, reporting relationships, and lines of communication. The applicant also should address the following:

- When more than one organization is responsible for HFE [such as instrumentation and control (I&C) and operations], the lead organizational unit answerable for the HFE program plan should be identified. If organization changes are expected over time (e.g., from design through construction to startup) necessary transitions between responsible organizations should be described.

- The team should have the authority and organizational placement to reasonably assure that all its areas of responsibility are completed, and to identify problems in establishing the overall plan or modifying its design.

- The team should have the authority to control further processing, delivery, installation, or use of HFE products until the disposition of a nonconformance, deficiency, or unsatisfactory condition is resolved.

(3) *Composition* – The applicant's HFE design team should include the expertise described in the appendix to this report.

(4) *Team Staffing* – The applicant should describe team staffing in terms of job descriptions and assignments of team personnel.

2.4.3 HFE Process and Procedures

(1) *General Process Procedures* – The applicant should identify the process through which the team will execute its responsibilities. It should include procedures for the following:

- assigning HFE activities to individual team members
- governing the internal management of the team
- making decisions on managing the HFE program
- making HFE design decisions
- controlling changes in design of equipment
- reviewing of HFE products

(2) *Process Management Tools* – The applicant should identify the tools and techniques (e.g., review forms) the team uses to verify that they fulfilled their responsibilities.

(3) *Integration of HFE and Other Plant or Modification Design Activities* – The applicant should describe the process for integrating the design activities (i.e., the inputs from other design work to the HFE program, and the outputs from the HFE program to other plant design activities). The applicant should also discuss the iterative aspects of the HFE design process.

(4) *HFE Program Milestones* – The applicant should identify HFE milestones that show the relationship of the elements of the HFE program to the integrated plant design, development, and licensing schedule. A relative program schedule of HFE tasks should be available for the NRC staff's review showing relationships between the HFE elements and the activities, products, and reviews.

Additional Information: A milestone might include, for example, the date when a simulator will be available for integrated system validation and operator training.

(5) *HFE Documentation* – The applicant should identify the HFE documentation items, such as RSRs and their supporting materials, and briefly describe them, along with the procedures for their retention and for making them available to the NRC staff for review.

(6)   *Subcontractor HFE Efforts* – The applicant should include HFE requirements in each subcontract contributing to the HFE program. The applicant should periodically verify the subcontractor's compliance with HFE requirements. The HFE plan should describe milestones and the methods used for this verification.

2.4.4   Tracking HFE Issues

(1)   *Availability* – The applicant should have a tracking system to address human factors issues that are:

- known to the industry (defined in the Operating Experience Review element, see Section 3)

- identified throughout the life cycle of the HFE aspects of design, development, and evaluation

- deemed by the HFE program as human engineering discrepancies (HEDs) (see Section 11.4.4)

Additional Information: Issues are those items that need to be addressed later, and hence must be tracked to assure that they are not overlooked. Establishing a new system to track HFE issues independent from the rest of the design effort is unnecessary; rather, an existing one can be adapted for this purpose (such as a plant's corrective-action program).

(2)   *Method* – The applicant's method should:

- establish criteria for when issues are entered into the system

- track issues until the potential for negative effects on human performance is reduced to an acceptable level.

(3)   *Documentation* – The applicant should document the actions taken to address each issue in the system; if no action is required, this should be justified.

Additional Information: The description of the final resolution of the issue should be sufficiently detailed so that a third party can understand how it was resolved.

(4)   *Responsibility* – After identifying an issue, the applicant's tracking procedures should describe individual responsibilities for logging, tracking, and resolving it, along with the acceptance of the outcome.

2.4.5   Technical Program

(1)   The applicant should describe the applicability and status of each of the following HFE elements:

- Operating Experience Review

- Functional Requirements Analysis and Function Allocation

- Task Analysis

- Staffing and Qualifications

- Treatment of Important Human Actions

13

- HSI Design
- Procedure Development (Described in SRP, Chapter 13 submittal)
- Training Development (Described SRP, Chapter 13 submittal)
- Human Factors Verification and Validation
- Design Implementation
- Human Performance Monitoring

Additional Information: The applicant should identify each applicable element of the HFE program. If the applicant determines that an HFE element is not applicable to the HFE program, the applicant should give a rationale. For example, if an applicant's HFE program involves modifying a control room HSI wherein the level of automation is not affected, then the Functional Requirements Analysis and Function Allocation element might not be included.

The applicant should describe the status of each element in the HFE plan (i.e., will the element be enacted in the future, is it currently being performed, or is it completed). The applicant should clearly identify the use of past analyses that the NRC has not reviewed (i.e., analyses originally undertaken for another design) and justify their use in the current application.

The criteria for the technical review of each element in the HFE program are presented in Sections 3 to 13 of this document.

(2)   The applicant should identify the approximate schedule for completing any HFE activities that are unfinished at the time of the application.

Additional Information: For example, if an applicant for design certification has not finished V&V, the applicant should give an approximate schedule for its completion.

(3)   The applicant's plan should identify and describe the standards and specifications that are sources of the HFE requirements.

(4)   The applicant's plan should specify HFE facilities, equipment, tools, and techniques (such as laboratories, simulators, rapid prototyping software) that the HFE program will employ.

(5)   *Additional Considerations for Reviewing the HFE Aspects of Plant Modifications* – The applicant should provide assurance that a modification to the control room or a change to risk-important human actions does not compromise defense in depth in accordance with RG 1.174. The applicant should assure the following important aspects of defense in depth:

- A reasonable balance is preserved among prevention of core damage, prevention of containment failure, and consequence mitigation.
- There is no over-reliance on programmatic activities to compensate for weaknesses in plant design. This may be pertinent to changes in credited human actions (HAs).
- System redundancy, independence, and diversity are preserved commensurate with the expected frequency, consequences of challenges to the system, and uncertainties (e.g., no risk outliers).

- Defenses against potential common cause failures are preserved, and the potential for the introduction of new common cause failure mechanisms is assessed. Caution should be exercised in crediting new HAs to verify that the possibility of significant common cause errors is not created.

- Independence of barriers is not degraded.

- Defenses against human errors are preserved. For example, establish procedures for a second check or independent verification for risk-important HAs to determine that they have been performed correctly.

- The intent of the General Design Criteria (GDC) in Appendix A to 10 CFR Part 50 is maintained. GDC that may be relevant are:
  - 3 - Fire Protection
  - 13 - Instrumentation and Control
  - 17 - Electric Power Systems
  - 19 - Control Room
  - 34 - Residual Heat Removal
  - 35 - Emergency Core Cooling System
  - 38 - Containment Heat Removal
  - 44 - Cooling Water.

- Safety margins often used in deterministic analyses to account for uncertainty and provide an added margin to provide adequate assurance that the various limits or criteria important to safety are not violated. Such safety margins are typically not related to HAs, but the reviewer should take note to see if there are any that may apply to the particular case under review. It is also possible to add a safety margin (if desired) to the HA by demonstrating that the action can be performed within some time interval (or margin) that is less than the time identified by the analysis.

Additional Information: Defense in depth, described in RG 1.174, is one of the fundamental principles upon which a plant is designed and built. It uses multiple means to assure safety functions and to prevent the release of radioactive materials. Defense in depth is important in accounting for uncertainties in equipment and human performance, and for ensuring some protection remains, even in the face of significant breakdowns in particular areas, such as safety systems, training, and quality assurance. Whereas an applicant may change a specific defense in depth strategy, defense in depth must be maintained overall. These types of defense in depth evaluations may be done as part of the 10 CFR 50.59 evaluation for modifying the plant.

2.5   Bibliography

IEC 60964: *Nuclear Power Plants-Control Rooms - Design* (International Electrochemical Commission, 2009).

IEEE Std. 1023-2004: *IEEE Recommended Practice for the Application of Human Factors Engineering to Systems, Equipment, and Facilities of Nuclear Power Generating Stations and Other Nuclear Facilities* (Institute of Electrical and Electronics Engineers, 2004).

ISO 11064-1: *Ergonomic Design of Control Centres -- Part 1: Principles for the Design of Control Centres* (International Standards Organization, 2000).

NUREG-0696: *Functional Criteria for Emergency Response Facilities* (NRC, 1981).

Regulatory Guide 1.174: *An approach for using probabilistic risk assessment in risk-informed decisions on plant-specific changes to the licensing basis* (NRC, 2011).

U.S. Code of Federal Regulations, Parts 20, 50, 52, and 55, Title 10, "Energy."

3 OPERATING EXPERIENCE REVIEW

3.1 Background

Applicants should provide the NRC staff with the administrative procedures that they will use for evaluating operating, design, and construction experience (referred to collectively as "operating experience"), and for ensuring that germane industry experiences will be provided in a timely manner to those designing and constructing the plant [10 CFR 50.34(f)(3)(i) and 52.47(a)(22)].

The main reason an applicant conducts an operating experience review (OER) as part of the HFE program is to identify HFE-related safety issues. The OER provides information on the past performance of predecessor designs. For new plants, these predecessors may be earlier designs upon which the new design is based. For modifications to plants, they may be the design of the systems being changed. The issues and lessons learned from operating experience offer a timely basis for improving the plant's design (i.e., at the beginning of the design process).

Considering an applicant's submittal for a new NPP, its predecessor designs are those plants, systems, HSIs, and operational approaches that are the basis for the new plant's design. It may be based on multiple predecessors and encompass both non-nuclear and nuclear industry sources.

An applicant's NPP design may be an evolutionary one, based on changes to an existing design that was used for several operating plants for many years. Alternatively, an applicant's new NPP design may be an innovative break from past designs that employs new technology or new operational approaches to realize improvements in safety, performance, availability, or reliability. More likely, such a new plant fits somewhere on the continuum between a traditional evolutionary plant and a completely novel design. All plants on this continuum, even innovative ones, may include certain systems and HSIs similar to those in existing operating plants, or evolved from them.

Regardless of where on this succession the applicant's design lies, it is vital to identify those plants, systems, HSIs, and operational approaches that are precursors to, or serve as the basis or departure point for the new design, so that their operating experience can be documented and evaluated to identify lessons learned.

The resolution of OER issues may involve changes in function allocation, automation, HSI equipment design, procedures, training, and so forth. Thus, the applicant can identify and analyze negative features in previous designs so that they are avoided in developing the current system while retaining the positive features.

OER information contributes inputs to other review elements, as summarized in Table 3-1. Thus, the OER can contribute to reviewing and evaluating system-design and other HFE considerations. For example, the OER can be used in selecting specific failure scenarios to incorporate in validation testing, and as a basis in choosing specific performance measures to evaluate (e.g., to measure an aspect of human performance identified as problematic in the OER).

Table 3-1 The Role of Operating Experience Review in the HFE Program

HFE Element	OER Contribution
Functional Requirements Analysis and Function Allocation	Basis for initial requirements
	Basis for initial allocations
	Identification of need for modifications
Task Analysis, Human Reliability Analysis, and Staffing/Qualifications	Important human actions and errors
	Problematic operations and tasks
	Instances of staffing shortfalls
Human-System Interface, Procedures, and Training Development	Trade study evaluations
	Potential design solutions
	Potential design issues
Human Factors Verification and Validation	Tasks to be evaluated
	Event and scenario selection
	Performance measure selection
	Issue resolution verification

3.2   Objective

The NRC staff uses the review criteria in this section to verify that the applicant has reviewed previous designs that are similar to the current one under review, and has identified, analyzed and addressed HFE-related problems to ensure that any negative features in the predecessor designs are avoided in the current design while retaining their positive features.

3.3   Applicant Products and Submittals

The product of the applicant's OER is a completed review of operating experience that consists of the items identified during the review, together with their resolutions for how the new design addresses the pertinent issues.

The applicant should provide either an IP or a completed RSR. If the applicant submits an IP, it should describe the methodology for conducting the OER. The NRC will review it using the criteria set out in Section 3.4 below. Then the applicant will submit the RSR when the work described by the IP is completed.

If the applicant submits a completed RSR, the NRC will verify the results using the criteria set out in Section 3.4 below. At a minimum, the RSR should include the following:

- identification of predecessor/related plants and systems

- methodology used to review the OE (may refer to an approved IP, if used)

- list of OE sources/documents reviewed

- discussion of the conduct of the OER, and of the results of reviewing relevant HSI technology

- description of, and findings from interviews with plant personnel or other users

- listing of OER-identified issues incorporated into the design

- enumeration of open issues still being tracked in the HFE issues-tracking system

Summaries may be used for any of the above items if references are given for more detailed documents. If the methodology was described in an IP that the NRC staff previously reviewed, the contents of the RSR should be consistent with the approved methodology and the applicant should discuss the rationale for any deviations from it.

In addition to evaluating the IP or RSR, the NRC staff reviewer also may audit the issue-tracking system at the applicant's facility to determine how OER issues were resolved.

3.4   Review Criteria

3.4.1   Scope

(1)   *Predecessor/Related Plants and Systems* – The applicant's OER should include information about human factors issues in the predecessor plant(s) or highly similar plants, systems, and HSIs, including the following:

- The OER should identify previous or predecessor design(s)/plant(s) used as part of the design basis of the plant being reviewed.

- The OER should define the relevance of each predecessor plant/design to the new design, when there is more than one predecessor.

- The OER should detail how the applicant identified and analyzed any HFE-related problems in the previous plants/designs, and how these issues are avoided in the new design.

- The OER should address how the applicant identified, evaluated, and incorporated or retained any positive features of previous plants/designs.

- The OER should describe the predecessor plant(s) and systems, explaining the relationship of each to the new design.

- For applicants proposing to use new technology or systems that were not used in the predecessor plants, the OER should review and describe the operating experience of any other facilities that already use that technology.

(2)   *Recognized Industry HFE Issues* – The applicant should address the HFE issues identified in NUREG/CR-6400. The issues are organized into the following categories:

- unresolved safety issues/generic safety issues (See 10 CFR 52.47(a)(21) and NUREG-0933)

- TMI issues

- NRC generic letters and information notices

- operating experience reports in the NUREG-1275 series, Vol. 1 through 14

- low power and shut down operations

- operating plant event reports

Additionally, the applicant should review and discuss all operating experience in the preceding categories that was published since NUREG/CR-6400 was published in 1996.

(3)   *Related HSI Technology* – The applicant's OER should cover operating experience with the proposed HSI technology in the applicant's design.

Additional Information: For example, if a computer operated support system, a computerized procedures system, or advanced automation are planned to be used, the OER should describe the HFE issues associated with using them.

(4)   *Issues Identified by Plant Personnel* – The applicant's OER should discuss issues identified through interviews with plant personnel based on their operating experience with plants or systems applicable to the new design. As a minimum, the interviews should include the following topics:

- Plant Operations
    - normal plant evolutions (e.g., startup, full power, and shutdown)
    - failure modes and degraded conditions of the I&C systems, including, but not limited to, the sensor, monitoring, automation and control, and communications subsystems. These include, for example, the safety-related system logic and control unit, fault tolerant controller (nuclear steam supply system), the local "field unit" for the multiplexer (MUX) system, the MUX controller (balance-of-plant), and a break in the MUX line failure modes
    - degraded conditions of the HSI resources (e.g., losses of video display units, of data processing, and of large overview display)
    - transients (e.g., turbine trip, loss of offsite power, station blackout, loss of all feedwater, loss of service water, loss of power to selected buses or MCR power supplies, and safety/relief valve transients)
    - accidents (e.g., main steam line break, positive reactivity addition, control rod insertion at power, control rod ejection, anticipated transients without scram, and various-sized loss-of-coolant accidents)
    - reactor shutdown and cooldown using the remote shutdown system

- HFE Design Topics
    - alarms and annunciation
    - displays
    - controls and automation
    - information processing and job aids
    - real-time communications with plant personnel and other organizations
    - procedures, training, staffing/qualifications, and job design

(5)   *Important Human Actions* – The applicant's OER should identify important HAs in the predecessor plants or systems (Section 7 defines important HAs), and determine whether they remain important in the applicant's design. Additional considerations cover the following:

- For the important HAs, the OER should identify the scenarios wherein actions are needed, and state whether they were needed and successfully completed. Those aspects of the design that helped ensure success should be identified.

- If errors occurred in the execution of the HAs, the applicant should identify insights to the needed improvements in human performance.

- When important HAs for the new plant are determined to differ from those of the predecessor plant, the OER should specify whether there is any operational experience with these different HAs.

3.4.2 Issue Analysis, Tracking, and Review

(1) *OER Process* – The applicant should discuss the administrative procedures for evaluating the operating, design, and construction experience, and for ensuring that applicable important industry experiences will be provided in a timely manner to those designing and constructing the plant.

Additional Information: 10 CFR 50.34(f)(3)(i) requires these administrative procedures.

(2) *Analysis Content* – The applicant should analyze issues to identify:

- human performance issues and sources of human error

- design elements supporting and enhancing human performance

(3) *Documentation* – The applicant should document the analysis of operating experience.

(4) *Incorporation Into the Tracking System* – The applicant should document each issue determined to be relevant to the design, but yet to be addressed, in the issue-tracking system (see Section 2.4.4).

3.4.3 Plant Modifications

(1) *Additional Considerations for Reviewing the HFE Aspects of Plant Modifications* – In addition to any of the criteria above that relate to the modification being reviewed, the applicant should address the following considerations:

- The focus of the scope of the applicant's OER should provide information on the plant's systems, HSIs, procedures, or training that are being modified.

- The applicant's OER should account for the operating experience of the plant that will be modified, including experiences with the systems that will be changed, and with technologies similar to those being considered.

Additional Information: Useful information may be found in the plant's corrective action program.

3.5   Bibliography

IAEA Safety Series No. 75-INSAG-3, Rev. 1: *Basic Safety Principles for Nuclear Power Plants* (International Atomic Energy Agency, 1999).

IEEE Std. 1023-2004: *IEEE Recommended Practice for the Application of Human Factors Engineering to Systems, Equipment, and Facilities of Nuclear Power Generating Stations and Other Nuclear Facilities* (Institute of Electrical and Electronics Engineers, 2004).

NUREG/CR-6400: *HFE Insights For Advanced Reactors Based Upon Operating Experience* (Higgins & Nasta, 1996).

4   FUNCTIONAL REQUIREMENTS ANALYSIS AND FUNCTION ALLOCATION

4.1   Background

Plant designers conduct functional requirements analysis (FRA) and function allocation (FA) to ensure that the functions necessary to accomplish plant goals are sufficiently defined and analyzed so that the allocation of functions to personnel and machine resources can take advantage of human and machine strengths and avoid human and machine limitations.

FRA is the identification of functions that must be performed to satisfy the nuclear power plant's overall goals:

- to ensure the health and safety of the public by preventing or mitigating the consequences of postulated accidents

- to generate power, i.e., supply electricity to the grid

The plant's goals are accomplished by high-level functions. The functions that address the plant's goal to ensure safety are often termed safety functions. Examples of safety functions are reactivity control, containment integrity and reactor coolant system water mass inventory control. These safety functions are often defined in terms of a boundary or parameter important to assuring the plant's integrity, and to preventing the release of radioactive materials. The safety function is often described without reference to specific plant systems and components, or the level of human and machine intervention needed to perform the function.

Applicants conduct an FRA to:

- define the high-level functions that have to be accomplished to meet the plant's goals and desired performance

- delineate the relationships between high-level functions and the plant's systems (e.g., plant configurations or success paths) responsible for performing the functions

- provide a framework for determining the roles and responsibilities of personnel and automation

Functions are essentially hierarchical (IEC, 2000). The term "function" can refer to high-level plant functions, such as safety functions, or to a lower-level description of the purpose of an individual piece of equipment, such as a valve or display system. Plants have a hierarchical structure of functions, processes, systems, and components. High-level functions are usually accomplished through some combination of lower-level system actuations such as reactor trip, safety injection, or accumulators. Often plant systems are used in combination to achieve a high-level function. The combination of systems used to achieve a high-level function is called a process (e.g., feed and bleed of the reactor coolant system). There may be more than one possible process that can achieve a given high-level function. Figure 4-1 provides a simple illustration of one part of the functional hierarchy associated with a commercial nuclear plant's goal to ensure safety.

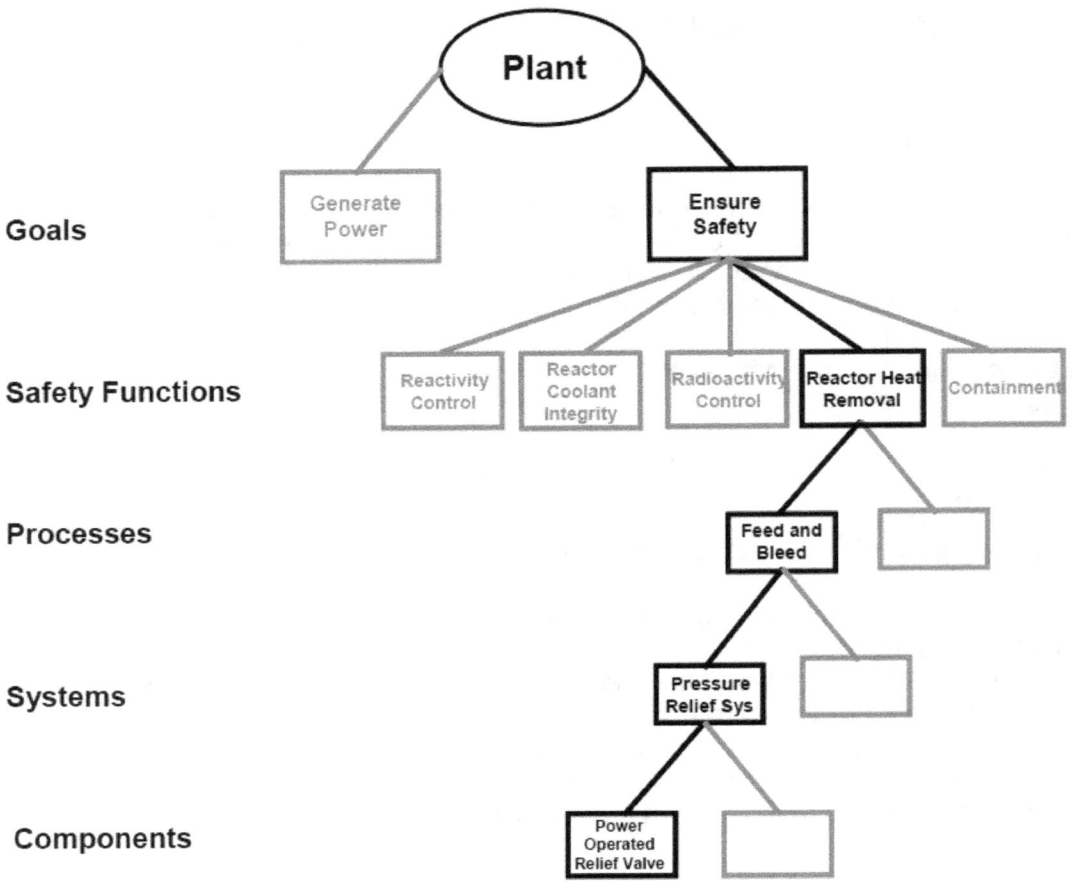

Figure 4-1 Vertical slice through a plant's functional hierarchy for ensuring safety

As functions are analyzed, their requirements become better defined. High-level functions can be broken down into the actions that are necessary to perform that function, whether those actions are performed by personnel or automation.

Once the functional requirements are understood, the designer assigns the functions to personnel and automation (hardware and software aspects of the plant), such as:

- personnel, e.g., manual control (no automation)
- automatic systems, e.g., fully automatic control, and passive, self-controlling phenomena
- combinations of personnel and automation, for example:
  - shared operation, the automatic operation of some aspects of a function, with others performed manually
  - operation by consent/delegation, automation takes control of a function when personnel direct it to do so under close monitoring and supervision
  - operation by exception, autonomous operation of a function, unless there are specific predefined situations or circumstances requiring manual human action

This assignment is called FA. Functions allocated to personnel are those that will be performed by HAs. Designers allocate functions to meet the functional requirements, while considering technology readiness and cost as well. To assure the safety and reliability of a design, designers also consider the relative capabilities, strengths, and weaknesses of personnel and automation. For example, if a functional requirement is that an action must be performed within seconds of a pump trip, it probably should be automated because operators are unlikely to recognize the trip and take the required action so quickly. In many cases, assuring the achievement of a control function requires allocating overlapping and redundant responsibilities to personnel and automation (e.g., assigning personnel the responsibility of monitoring and maintaining supervisory control over automated systems).

The FA process is where the role of personnel is initially defined. The role of personnel is essentially the aggregate of all HAs across functions. Applicants further analyze HAs to identify the specific tasks that personnel must perform to accomplish them. This is referred to as "Task Analysis" (see Section 5 of this document). Then tasks are assigned to specific staff positions to create jobs as part of Staffing and Qualifications analyses (see Section 6).

FRA and FA decisions are not limited to new plants; they also are important for plant modifications that may change the level of automation of the original design (e.g., to a plant's feedwater control system), which may affect the roles and responsibilities of plant personnel. Modifications may not only alter the tasks that personnel undertake when interacting with a new system or HSI, but may also impact how effectively they can perform other functions that may seem unrelated to the modification.

4.2   Objective

The NRC staff uses the review criteria in this section to verify that the applicant has:

- defined those functions that must be carried out to satisfy the plant's safety goals and its goal of generating power
- allocated those functions to personnel and automation in a way that takes advantage of human strengths and avoids human limitations

4.3   Applicant Products and Submittals

The product of the applicant's FRA is the complete set of functional requirements necessary to satisfy the plant's goals. The product of the FA is the identification of how personnel and automatic systems perform the functions.

The applicant should provide either an IP or a completed RSR. If the applicant submits an IP, it should describe the methodology for conducting the FRA and FA. The NRC will review it using the criteria set out in Section 4.4 below. Then the applicant will submit the RSR when the work described by the IP is completed.

If the applicant submits a completed RSR, the NRC will verify the results using the criteria set out in Section 4.4 below. At a minimum, the RSR should include the following:

- an explanation of the methodology used to define the safety functions
- the set of safety functions for the facility

- an explanation of the methodology used to allocate functions and the final set of allocations
- the technical basis for modifying high-level functions of predecessor plants in the new design
- a complete set of functional requirements necessary to satisfy the plant goals
- identification of how personnel and automatic systems perform the functions
- the technical basis for all function allocations

Summaries may be used for any of the above items provided that references are given for more detailed documents. If the methodology was described in an IP that the NRC staff previously reviewed, the contents of the RSR should be consistent with the approved methodology and the applicant should discuss the rationale for any deviations from it.

4.4   Review Criteria

(1)   The applicant should use a structured, documented methodology reflecting HFE principles to perform functional requirements analysis (FRA) and function allocation (FA).

Additional Information: Figure 4-2 is an example of an FRA and FA process. The FRA and FA may be graded based on:

- the degree to which the functions of the new design differ from those of its predecessor(s)
- the extent to which problems in operating experience were encountered for the plant's functions in predecessor plants

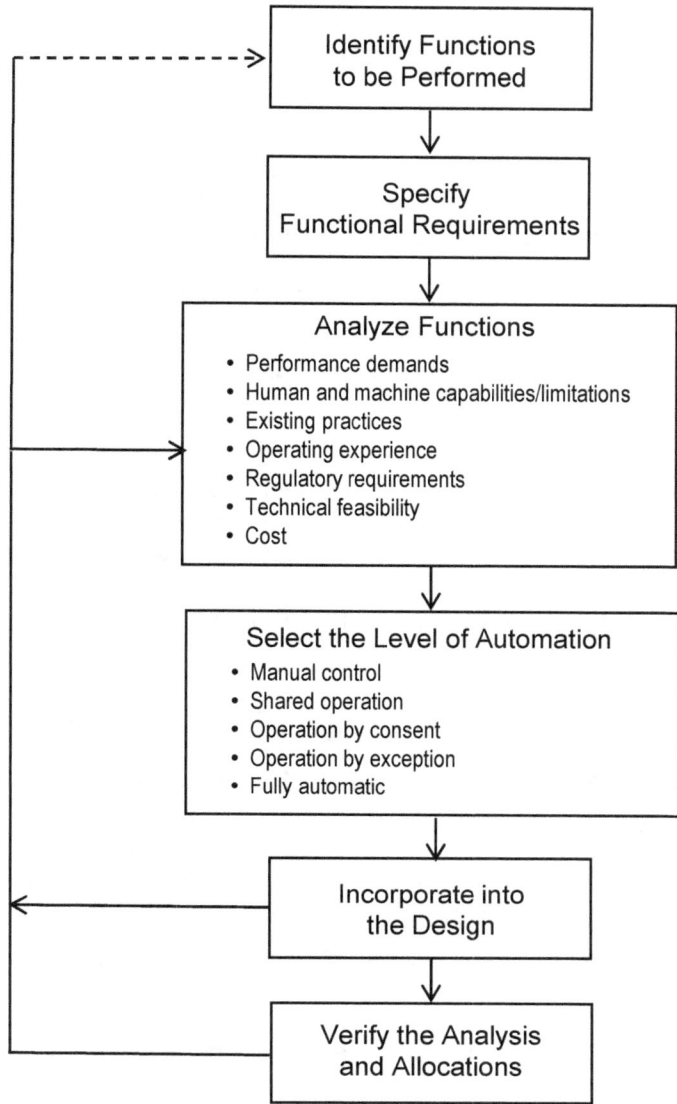

Figure 4-2 Allocation of functions to personnel and automatic systems

(2)   The applicant's FRA and FA should be performed iteratively to keep it current during design development and operation up to decommissioning, so that it can be used as a design basis when modifications are considered.

(3)   The applicant should describe the plant's functional hierarchy, including, as appropriate goals, functions, processes, and systems. The description should include:

- comparing them with the predecessor or reference plants and systems, i.e., the previous ones on which the new plant is based

- identifying the differences between the proposed and reference plants and systems

- documenting the technical basis for modifications to high-level functions in the new design (compared to the predecessor design)

- defining, for each safety function and other plant function (e.g., electrical power generation), the set of system configurations or success paths that are responsible for, or able to carry out the function

- decomposing the functions, starting at "high-level" functions where a very general picture of major functions is described, and continuing to lower levels, until a specific critical end-item requirement emerges (e.g., a piece of equipment, software, or an HA). The functional decomposition should address the following levels

  - high-level functions (e.g., maintain reactor coolant system integrity)

  - the processes, as appropriate, that enable achievement of these functions.

  - specific plant systems and components

  - HAs, as appropriate

Additional Information: Safety functions (e.g., reactivity control) include functions needed to prevent or mitigate the consequences of postulated accidents that could pose undue risk to the public's health and safety. HAs will be further evaluated in the task analyses.

(4)   For each high-level function, the applicant should identify requirements related to:

- purpose of the high-level function

- conditions indicating that the high-level function is needed

- parameters indicating that the high-level function is available

- parameters indicating that the high-level function is operating (e.g., flow indication)

- parameters indicating that the high-level function is achieving its purpose (e.g., reactor vessel level returning to normal)

- parameters indicating that the operation of the high-level function can or should be terminated

Additional Information: At this stage, parameters may be described qualitatively (e.g., high or low). Specific data values or setpoints are not necessary.

(5)   Applicants should allocate functions to a level of automation (e.g., from manual to fully automatic) and identify the technical bases for the allocations.

Additional Information: The technical basis for the FA can be any one or combination of the factors (see Figure 4-2). For example:

- Functions, or parts of them, may be allocated based on operating experience. Successful operating experience may suggest keeping allocations the same as in predecessor designs and operating experience issues may suggest changing the allocations to address the issues.

- Functions, or parts of them, may be allocated to automation when their performance requirements exceed human capabilities and human error is likely. Conditions that establish a basis for automation (assuming the acceptability of other factors, such as technical feasibility or cost) include when the required response time is very short, when an action has to be performed repeatedly, or when very precise control is required.

- Functions, or parts of them, should be allocated to personnel when human knowledge and judgment is needed to ensure reliable function performance, it is important to keep personnel involved in the actions so they have good situation awareness should they need to perform the function, or to preclude boredom.

(6) The applicant's FA should consider not only the primary allocations to personnel (those functions for which personnel have the primary responsibility), but also their responsibilities to monitor automatic functions, detect degradations and failures, and to assume manual control when necessary.

(7) The applicant should describe the overall role of personnel by considering all functions allocated to them.

Additional Information: The FA to personnel and automation is considered on a function-by-function basis. However, the overall personnel role is an aggregate of all functions allocated to them. While on an individual basis, a single function allocation to personnel may be justified, allocations should also be considered in the context of other responsibilities personnel have to help ensure that together all functions allocated to personnel are acceptable and do not interfere with each other.

(8) The applicant should verify that the FRA and FA accomplish the following:

- all the high-level functions needed to achieve safe operation are identified
- all requirements of each high-level function are identified
- the allocation of functions to humans and automatic systems assures a role for personnel that takes advantage of human strengths and avoids human limitations

(9) *Additional Considerations for Reviewing the HFE Aspects of Plant Modifications* - In addition to any of the criteria above that relate to the modification being reviewed, the applicant should address the following considerations:

- The FRA should address new functions resulting from changes in the degree of integration between plant systems.

  Additional Information: The FRA for modifications may change existing safety functions or introduce new functions for systems supporting them. For example, installing higher-level automation may bring systems formerly controlled separately under a single controller. Also, the modifications may change the degree to which different plant systems share common resources (e.g., power, cooling water, and data-transmission buses). These may be important in diagnosing malfunctions or planning responses.

- The FRA should be revised and updated to reflect the modification. The scope of the FRA may be restricted to functions related to the modification.

- The FA should be revised and updated to reflect modifications that are likely to change the allocation between personnel and plant systems of functions important to safety. The scope of the analyses may be restricted to functions involving the modification.

- A change in the role of personnel due to a modification should be examined within the context of its effects on their overall responsibilities.

Additional Information: Increases in certain task demands may affect the ability of personnel to carry out other actions categorized as important (see Section 7).

4.5   Bibliography

ANSI/ANS 58.8: *Time Response Design Criteria for Safety-Related Operator Actions* (American Nuclear Society, 1994).

BNL Technical Report 91017-2010: *Human-System Interfaces to Automatic Systems: Review Guidance and Technical Basis* (O'Hara & Higgins, 2010).

IAEA-TECDOC-668: *The Role of Automation and Humans in Nuclear Power Plants* (International Atomic Energy Agency, 1992).

IEC 60964: *Nuclear Power Plants-Control Rooms - Design* (International Electrochemical Commission, 2009).

IEC 61839: *Nuclear Power Plants - Design of Control Rooms - Functional Analysis and Assignment* (International Electrochemical Commission, 2000)

NUREG/CR-2623: *The Allocation of Functions in Man-Machine Systems: A Perspective and Literature Review* (Price, et al., 1982).

NUREG/CR-3331: *A Methodology for Allocation of Nuclear Power Plant Control Functions to Human and Automated Control* (Pulliam et al., 1983).

5   TASK ANALYSIS

5.1   Background

As noted in Section 4, the functions allocated to personnel are those that will be performed by HAs. Applicants further analyze HAs to identify the tasks that personnel must perform to accomplish them. Tasks are a group of related activities with a common objective. Task analysis identifies the specific tasks needed to accomplish HAs, and the information, control, and task support required to complete those tasks. Accordingly, the results of task analyses are identified as inputs in many HFE activities; e.g., they form the basis for evaluating:

- staffing and qualifications
- HSIs, procedures, and training program design
- task support verification (see Human Factors Verification and Validation in Section 11)

5.2   Objective

The NRC staff uses the review criteria in this section to verify that the applicant has performed analyses that:

- identify the specific tasks personnel perform to accomplish their functions
- identify the alarms, information, controls, and task support needed to perform those tasks

5.3   Applicant Products and Submittals

The product of the applicant's task analysis is a listing of the tasks to be undertaken, and the requirements for performing each task. The task analysis generates inputs to other elements in the HFE process.

The applicant should provide either an IP or a completed RSR. If the applicant submits an IP, it should describe the methodology for conducting the task analysis. The NRC will review it using the criteria set out in Section 5.4 below. Then the applicant will submit the RSR when the work described by the IP is completed.

If the applicant submits a completed RSR, the NRC will verify the results using the criteria in Section 5.4 below. At a minimum, the RSR should include the following:

- the HAs to be addressed by task analysis
- a description of the task analysis methodology
- a description of
    - personnel tasks including a narrative of the activities to be perform
    - the applicable aspects of the tasks (see Criterion 2 below)
    - the relationship between tasks
    - an estimate of the time needed to perform the tasks
    - estimated workload

- a list of the alarms, information, controls, and task support identified by task analysis
- an identification of the number of personnel needed to complete each task
- a designation of the knowledge and abilities needed to perform each task

Summaries may be used for any of the above items provided that references are given for more detailed documents. If the methodology was described in an IP that the NRC staff previously reviewed, the contents of the RSR should be consistent with the approved methodology and the applicant should discuss the rationale for any deviations from it.

5.4   Review Criteria

(1)   The scope of the applicant's task analysis should include:

- All important HAs as determined by probabilistic and deterministic means (see Section 7, Treatment of Important Human Actions, of this report)
- The applicant should select tasks for analysis that represent the full range of plant operating modes, including startup, normal operations, low-power and shutdown conditions, transient conditions, abnormal conditions, emergency conditions, and severe accident conditions. The chosen tasks should cover:
    - tasks that were not identified as "important HAs" but have negative consequences if performed incorrectly
    - tasks that are new compared to those in predecessor plants, such as ones related to new systems or procedures
    - tasks that, while not new, are performed significantly differently from predecessor plants
    - tasks related to monitoring of automated systems that are important to plant safety, and the use of automated support aids for personnel, such as computer-based procedures
    - tasks related to identifying the failure or degradation of automation, and implementing backup responses
    - tasks anticipated to impose high demands on personnel, e.g., little time or high workload (such as administrative tasks that contribute to work load and challenge ability to monitor the plant)
    - tasks important to plant safety that are undertaken during maintenance, tests, inspections, and surveillances
    - tasks with potential concerns for personnel safety (such as maintenance tasks performed in the containment)

(2)   The applicant should describe the screening methodology used to select the tasks for analysis, based on criteria specifically established to determine whether analyzing a particular task is necessary.

(3)   The applicant should begin task analysis with detailed narratives of what personnel have to do. The analysis should be sufficiently detailed to define the alarms, information, controls, and task support needed to accomplish the task. The detailed task descriptions should address (as applicable to the task) the topics listed in Table 5-1.

Table 5-1 Task Considerations

Topic	Example
Alerts	• alarms and warnings
Information	• parameters (units, precision, and accuracy) • feedback needed to indicate adequacy of actions taken
Decision-making	• decision type (relative, absolute, probabilistic) • evaluations to be performed

Response	• actions to be taken
	• task frequency and required accuracy
	• time available and temporal constraints (task ordering)
	• physical position (stand, sit, squat, etc.)
	• biomechanics
	    - movements (lift, push, turn, pull, crank, etc.)
	    - forces needed

Teamwork and Communication	• coordination needed between the team performing the work • personnel communication for monitoring information or taking control actions
Workload	• cognitive • physical • overlap of task requirements (serial vs. parallel task elements)
Task Support	• special and protective clothing • job aids, procedures or reference materials needed • tools and equipment needed
Workplace Factors	• ingress and egress paths to the worksite • workspace needed to perform the task • typical environmental conditions (such as lighting, temp, noise)
Situational and Performance Shaping Factors	• stress • time pressure • extreme environmental conditions • reduced staffing
Hazard Identification	• identification of hazards involved, e.g., potential personal injury

(4)   The applicant should identify the relationships among tasks.

Additional Information: For example, some tasks can be carried out in any order or in parallel, some tasks have to be performed in a linear sequence, while for others the relationship is conditional (if such a condition exists, perform task A). Some tasks may involve coordinated actions among crew members or control room crew members and local personnel.

(5)   The applicant should estimate the time required to perform each task.

(6)   The applicant should identify the number of people required to perform each task.

(7)   The applicant should identify the knowledge and abilities required to perform each task.

(8)   The applicant's task analysis should be iterative, and updated as the design is better defined.

(9)   Applicants should provide an analyses of the feasibility and reliability for important HAs that address the following:

- The analysis establishes the time available using an analysis method and acceptance criteria consistent with the regulatory guidance associated with the actions. The basis for the time available is documented.

  Additional information: The time available to perform the actions should be based on analysis of the plant response to the anticipated operational occurrence or accident. This analysis should reflect the guidance associated with the event.

- The analysis of the time required is based on a documented sequence of operator actions (based on task analysis, vendor-provided generic technical guidelines for emergency operating procedure development, or plant-specific EOPs, depending on the maturity of the design).

- Techniques to minimize bias are used when estimates of time required are derived using methods that are dependent on expert judgment. Uncertainties in the analysis of time required are identified and assessed.

- The sequence of actions uses only alarms, controls, and displays that would be available and operable during the assumed scenario(s).

- The estimated time for operators to complete the credited action is sufficient to allow successful execution of applicable steps in the EOPs.

  Additional Information: Acceptable methods for deriving analysis time estimates for individual task components include, but are not limited to:

  - Operator interviews and surveys
  - Operating experience reviews
  - Software models of human behavior, such as task network modeling
  - Use of control/display mockups
  - Expert panel elicitation (e. g., Kolaczkowski et al., 2007)

- Staffing for analysis is justified, and if credited manual actions require additional operators beyond the assumed staffing, the justification for timely availability of the additional staffing is provided and the estimate of time required includes any time needed for calling in additional personnel.

- The analysis of the action sequence is conducted at a level of detail sufficient to identify individual task components, including cognitive elements such as diagnosis and selection of appropriate response.

  Additional information: The documented sequence of operator actions should be analyzed at a level of detail necessary to identify critical elements of the actions and performance shaping factors (e.g., workload, time pressure) that affect time required and likelihood of successful completion of the action sequence. The applicant should establish time estimates for individual task components (e.g., acknowledging an alarm, selecting a procedure, verifying that a valve is open, starting a pump) and the basis for the estimates, through a method applicable to the HSI characteristics of digital computer-based I&C.

- The analysis identifies a time margin to be added to the time required and the basis for the adequacy of the margin.

(10) *Additional Considerations for Reviewing the HFE Aspects of Plant Modifications* – In addition to any of the criteria above that relate to the modification being reviewed and will affect HAs previously identified as important, cause existing ones to become important, or create new HAs that are important, the applicant should address the following considerations:

- Existing task analysis should be revised and updated to reflect the modification. If no pertinent task analysis exists, then consideration should be given to completing a new task analysis. For maintenance, tests, inspections, and surveillances, attention should be given to new important human actions, or those supported by new technologies (e.g., new capabilities for on-line maintenance).

- The task analysis should identify the design characteristics of the existing HSIs supporting the performance of experienced personnel (e.g., support high levels of performance during demanding situations) and consider them in developing new design requirements. Also, design features identified during the OER also should be carefully weighed in these analyses.

  Additional Information: The design characteristics may include the spatial arrangement of control and display devices and the ease of adjusting controls and displays to deal with special tasks. The new design should have features performing similar functions as the previous design, or should eliminate the need for the same features by performing these functions differently (e.g., by automating them). In addition, the task analysis should identify and examine any adjustments made to the previous HSIs by users, such as notes and external memory aids, suggesting that the previous design does not fully meet the users' needs. The new design requirements should adequately address all task demands.

5.5   Bibliography

A Guide to Task Analysis (Kirwan & Ainsworth, 1992).

Cognitive Task Analysis (Shraagen, Chipman & Shalin, 2000).

Cognitive Work Analysis: Toward Safe, Productive, and Healthy Computer-Based Work (Vicente, 1999).

IEC 60964: *Nuclear Power Plants-Control Rooms - Design* (International Electrochemical Commission, 2009).

IEC 61839: *Nuclear Power Plants-Design of Control Rooms-Functional Analysis and Assignment (IEC, 2000).*

NUREG-1852: *Demonstrating the Feasibility and Reliability of Operator Manual Actions in Response to Fire* (Kolaczkowski,et al., 2007).

NUREG/CR-3371: *Task Analysis of Nuclear Power Plant Control Room Crews* (Burgy et al., 1983).

6   STAFFING AND QUALIFICATIONS

6.1   Background

Plant staff and their qualifications are important considerations throughout the design process. Initial staffing levels may be established based on experience with previous plants, staffing goals (such as for staffing reductions), initial analyses, and government regulations. Final staffing levels result from the analyses described in this section, the applicant's policy and practices, and regulatory information.

Staffing levels are also important when plant modifications are designed. For example, when modifications impact important HAs, the applicant may review staffing needs to assure that those actions can be successfully accomplished. Many tasks require teamwork and communication between control-room staff, auxiliary operators, and other plant personnel. The NRC reviews the analyses the applicant employed to determine staffing qualifications and staffing levels for operating the facility and for completing the identified tasks. As another example, in proposing a modernization that significantly changes the technology underlying control room operations, the applicant should evaluate the effect of the change on the qualifications of the plant's staff. This element is used to review the applicant's staffing analyses for such modifications.

6.2   Objective

The NRC staff uses the review criteria in this section to verify that the applicant has systematically analyzed the required number and necessary qualifications of personnel, in concert with task requirements, and regulatory requirements.

The scope of applicable plant personnel are listed in Section 2.4.1, General HFE Program Goals and Scope, Criterion (6).

6.3   Applicant Products and Submittals

The product of the applicant's staffing and qualifications analyses defines the operating staff levels and the related qualification requirements for the particular facility.

The applicant should provide either an IP or a completed RSR. If the applicant submits an IP, it should describe the methodology for conducting staffing and qualifications analysis. The NRC will review it using the criteria in Section 6.4 below. Then the applicant will submit the RSR when the work described by the IP is completed.

If the applicant submits a completed RSR, the NRC will verify the results using the criteria in Section 6.4 below. At a minimum, the RSR should include the following:

- a description of the process used to determine initial and final staffing levels and personnel qualifications
- initial and final staffing levels
- the assignment of tasks to personnel
- a description of necessary qualifications of personnel

- input to the staffing evaluation from the other pertinent HFE elements, or a justification as to why no input was included

- results of validating the final staffing levels

Summaries may be used for any of the above items provided that references are given for more detailed documents. If the methodology was described in an IP that the NRC staff previously reviewed, the contents of the RSR should be consistent with the approved methodology and the applicant should discuss the rationale for any deviations from it.

6.4   Review Criteria

(1)   The applicant should address the applicable staffing and qualifications guidance in NUREG-0800 Section 13.1.

Additional Information: The NRC's reviewers for Chapter 18 of NUREG-0800 should verify that the reviews of Section 13.1 were completed.

(2)   The applicant should address the applicable staffing and qualifications guidance in 10 CFR 50.54.

Additional Information: As part of their verification, the Chapter 18 reviewers should assure that staffing meets the requirements of 10 CFR 50.54. For plant staffing levels that require an exemption from 10 CFR 50.54, the NRC's reviewers should use the guidance in NUREG-1791 (Persensky et al., 2005) and NUREG/CR-6838 (Plott et al., 2004).

(3)   The applicant should use the results of the task analysis as an input to the staffing and qualification analyses. Personnel tasks, addressed in task analysis, should be assigned to staffing positions to ensure that jobs are defined considering:

- the task characteristics, such as the knowledge and abilities required, relationships among tasks, time required to perform the task, and estimated workload

- the person's ability to maintain situation awareness within the area of assigned responsibility

- teamwork and team processes, such as peer checking

(4)   The applicant's staffing analysis should determine the number and qualifications of operations personnel for the full range of plant conditions and tasks, including operational tasks (under normal, abnormal, and emergency conditions), plant maintenance, plant surveillance, and testing.

Additional Information: The staffing analysis should address how the activities performed by personnel listed in Section 2.4.1, General HFE Program Goals and Scope, Criterion (5) impact and/or interface with the MCR. A reasonable approach is using predecessor plant data as a starting point for the analysis and adjusting the staffing numbers in accord with information from the new plant's design.

(5)   The applicant's staffing analysis should be iterative; that is, the initial staffing goals should be modified as information from the HFE analyses from other elements becomes available.

(6)   The applicant should address the basis for staffing and qualification levels considering the specific staffing-related issues noted below. These considerations may be identified in other HFE elements or in related source documents as follows:

- Operating Experience Review

    - operational problems and strengths resulting from staffing levels in predecessor designs

    - initial staffing goals and their bases, including staffing levels of predecessor designs and a description of significant similarities and differences between predecessor and current designs

    - staffing considerations described in NRC Information Notice 95-48, "Results of Shift Staffing Study"

    - possible impact on staffing of requirements of limits to work hours, required break times, and required days off, as specified in 10 CFR 26.205, Work Hours, as part of the Fitness for Duty Rule

    - Regulatory Issue Summary (RIS) 2009-10, Communications Between the NRC and Reactor Licensees During Emergencies and Significant Events

- Functional Requirements Analysis and Function Allocation

    - potential mismatches between functions allocated to personnel and their qualifications

    - changes to the roles of personnel due to modifying the plant's systems and HFE aspects

- Task Analysis

    - time needed to perform a task, and the workload involved

    - personnel communication and coordination, including interactions between individuals for diagnosing, planning, and controlling the plant, and interactions between personnel for administrative, communications, and reporting activities

    - the job requirements resulting from the sum of all tasks allocated to each individual inside and outside the control room

    - potential decreases in the ability of personnel to coordinate their work due to changes to the plant

    - availability of personnel considering other work that may be ongoing, and for which operators may be responsible outside the control room (e.g., fire brigade)

    - actions identified in 10 CFR 50.47, NUREG-0654, and procedures to implement an initial accident response in key functional areas, as denoted in the emergency plan

    - staffing considerations described by the application of ANSI/ANS 58.8-1994, "Time Response Design Criteria for Safety-Related Operator Actions" (ANS, 1994), if used by the applicant

- Treatment of Important Human Actions

    - the effect of staffing levels on the performance of the identified important HAs

    - the effect of staffing levels on personnel coordination for important HAs

- NUREG/CR-6753, Review of Findings for Human Performance Contribution to Risk in Operating Events
- Procedure Development
  - staffing demands resulting from requirements to concurrently use multiple procedures
  - personnel knowledge, abilities, and authorities identified in the procedures
- Training Program Development
  - concerns about coordinating personnel that are identified during the development of training

6.5   Bibliography

10 CFR 50.47: *U.S. Code of Federal Regulations*, Part 50, "Domestic Licensing of Production and Utilization Facilities," Title 10, "Energy."

10 CFR 50.54: *U.S. Code of Federal Regulations*, Part 50, "Domestic Licensing of Production and Utilization Facilities," Title 10, "Energy."

10 CFR 26: *U.S. Code of Federal Regulations*, Part 26, "Fitness for Duty Programs," Title 10, "Energy."

ANSI/ANS 58.8: *Time Response Design Criteria for Safety-Related Operator Actions* (American Nuclear Society, 1994).

ANSI/ANS 3.1: *Selection, Qualification, and Training of Personnel for Nuclear Power Plants*, (American Nuclear Society, 1993; R 1999).

ANSI/ANS 3.5: *Nuclear Power Plant Simulators for Use in Operator Training* (American Nuclear Society, 2009).

Generic Letter No. 82-12: *Policy on Factors Causing Fatigue of Operating Personnel at Nuclear Reactors* (NRC, 1982).

Information Notice 95-48: *Results of Shift Staffing Study* (NRC, 1995).

Information Notice 97-78: *Crediting of Operator Actions in Place of Automatic Actions and Modifications of Operator Actions, Including Response Times* (NRC, 1997).

NUREG-0654, Rev. 1: *Criteria for Preparation and Evaluation of Radiological Emergency Response Plans and Preparedness in Support of Nuclear Power Plants* (NRC, 1980).

NUREG-0737 and Supplements: *Clarification of TMI Action Plan Requirements* (NRC, 1980).

NUREG-0800: *Standard Review Plan*, Sections 13.1.1 - 13.1.3 (NRC, 2007).

NUREG-1122: *Knowledge & Abilities Catalogue for NPP Operators: PWRs* (NRC, 1998).

NUREG-1123: *Knowledge & Abilities Catalogue for NPP Operators: BWRs* (NRC, 2007).

NUREG-1791: *Guidance for Assessing Exemption Requests from the Nuclear Power Plant Licensed Operator Staffing Requirements Specified in 10 CFR 50.54(m)* (Persensky, Szabo, A., Plott, Engh & Barnes, 2005).

NUREG/CR-6753: *Review of Findings for Human Performance Contribution to Risk in Operating Events* (Gertman, et al., 2001).

NUREG/CR-6838: *Technical Basis for Regulatory Guidance for Assessing Exemption Requests from Nuclear Power Plant Licensed Operator Staffing Requirements Specified in 10 CFR 50.54(m)* (Plott, Engh & Barnes, 2004).

Regulatory Guide 1.8, Rev. 3: *Personnel Selection and Training* (NRC, 2000).

Regulatory Guide 1.114, Rev. 3: *Guidance to Operators at the Controls and to Senior Operators in the Control Room of a Nuclear Power Unit* (NRC, 2008).

Regulatory Guide 1.149, Rev. 4: *Nuclear Power Plant Simulators for Use in Operator Training* (NRC, 2011).

RIS 2009-10*: Communications between the NRC and Reactor Licensees during Emergencies and Significant Events* (NRC, 2009).

7   TREATMENT OF IMPORTANT HUMAN ACTIONS

7.1   Background

A goal of the NRC safety programs over the last several decades has been to use risk analyses to prioritize activities and to ensure that both regulators and applicants focus their efforts and resources on those activities that best assure the public's health and safety. HFE programs contribute to this goal by applying a graded approach to plant design by focusing greater attention to those HAs most important to safety.

Consequently, applicants identify those HAs most important to safety via a combination of probabilistic and deterministic analyses, and then address them when conducting the HFE program. The former typically is done using a probabilistic safety assessment or probabilistic risk assessment (PRA), including its human reliability analysis (HRA). These analyses identify the risk-important HAs described in Chapter 19 of the FSAR and in the DCD. Deterministic engineering analyses generally are completed as part of the suite of analyses in the FSAR/DCD in Chapters 7, Instrumentation and Controls, and 15, Transient and Accident Analyses. These analyses sometimes include credit for HAs by operators as part of an evaluation. Thus, a full identification of important HAs depends on analyses and methods that are reviewed by the NRC's staff using Chapters 7, 15, and 19 of the *Standard Review Plan* (NUREG-0800). This element offers more specific guidance related to these important HAs. We describe probabilistic analyses first, followed by a discussion of deterministic analyses.

HRA is an integral part of a completed PRA. Applicants submit PRAs in accordance with the NRC's current requirements. An HRA evaluates the potential for, and mechanisms of human error that might affect plant safety. Thus, it is an essential feature in assuring the HFE program goal of generating a design to minimize personnel errors, support their detection, and ensure recovery capability. The HRA is an integrated activity supporting both the HFE design and PRA activities. The robustness and quality of the HRA largely depends on the analyst's understanding of the causes, modes and probabilities of human error, the personnel tasks to be performed, information about those tasks, and any task-specific factors that may influence the human performance of them. Analysts should employ the descriptions and analyses of personnel functions and tasks, along with the operational characteristics of the HSIs. The HRA provides valuable insights into the desirable characteristics of the HSI design. Consequently, the HFE design should pay special attention to those plant scenarios, risk-important HAs, and HSIs that the PRA/HRA highlights as vital to plant safety and reliability.

The PRA and HRA should begin early in the design process to provide insights and guidance for both systems design and for HFE purposes. Thus, the applicant should use, as appropriate, the first version of the PRA/HRA (depending on the amount of design information available) to identify the important HAs, so that they can be considered in the early HFE design elements. The analyses should be updated iteratively as the design progresses (including the final PRA/HRA) to ensure the actual important HAs are captured and considered. At the very least, the initial PRA/HRA, and the set of important HAs, should be finalized when the design of the plant and HSI are complete.

Probabilistic analyses are supplemented by identifying important HAs in the FSAR/DCD deterministic analyses. To establish a licensing basis, applicants must analyze transients and accidents in accord with the requirements of 10 CFR 50.34 and 10 CFR 50.46; these events are described in the Standard Review Plan. The analyses appear in Chapter 15 of a DCD, or an

FSAR and in some cases include HAs that are credited in the analyses to prevent or mitigate the accidents and transients. These HAs may, or may not, be found as risk-important by the PRA. Nonetheless, all credited HAs should be considered deterministically as significant for the purposes of the HFE program.

The NRC I&C staff has established a position on common cause failures of digital I&C in a nuclear power plant (currently in the Interim Staff Guidance on *Diversity and Defense in Depth (D3) Issues* - NRC, 2009). Applicants are to perform a D3 analysis to demonstrate that their designs adequately address vulnerabilities to common cause failures. The applicant may identify backup systems or HAs necessary for accomplishing the required safety functions. These HAs should be treated as important human actions in the HFE program.

Figure 7-1 illustrates the relationship between the treatment of important HAs and the rest of the HFE program, as specified in Section 7.4, Criterion 3 below.

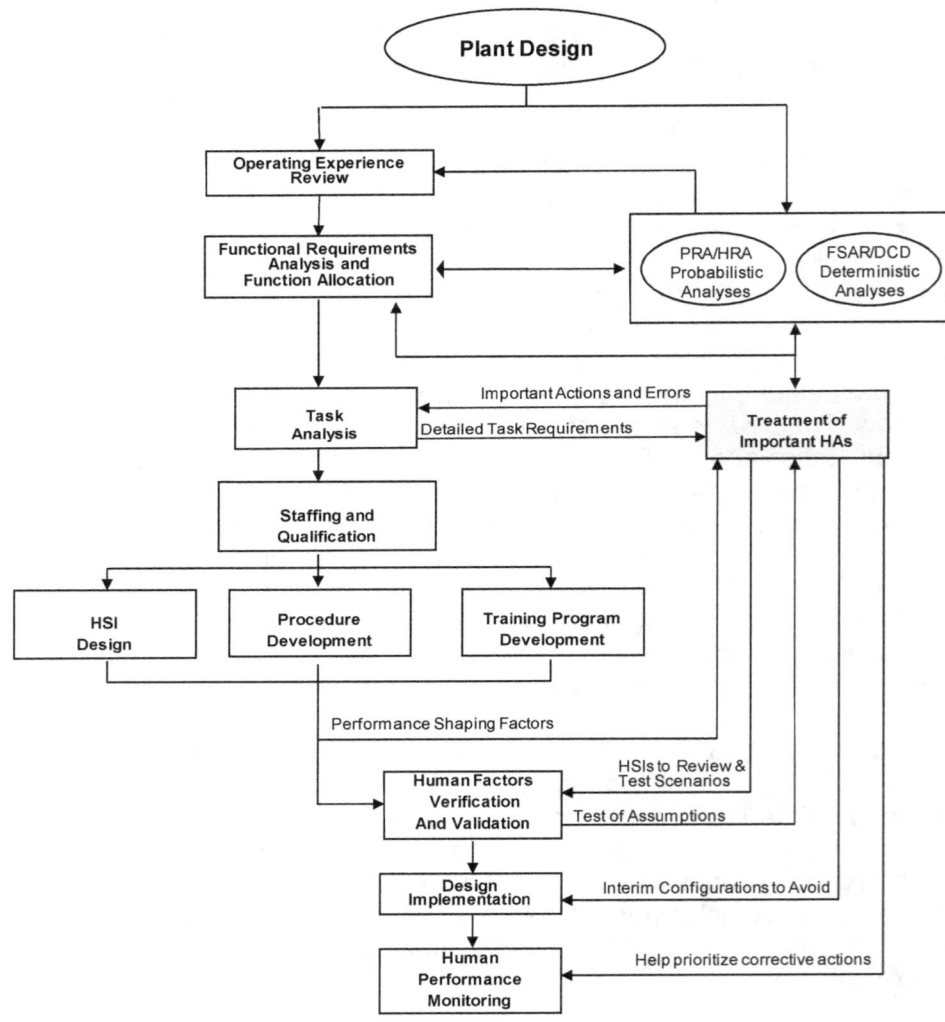

Figure 7-1 The role of important human actions in the HFE program

The important HAs are specifically addressed in many HFE elements, where the applicant describes how each of the important HAs is addressed in the HFE program.

7.2   Objective

The NRC staff uses the review criteria in this section to verify that the applicant has:

- identified the important HAs

- considered them in designing the HFE aspects of the plant to minimize the likelihood of personnel error, and to help ensure that personnel can detect and recover from any errors that occur

7.3   Applicant Products and Submittals

The product of the applicant's analysis of important human actions is their list of important HAs transferred to the other pertinent HFE elements, as noted in Section 7.3, Criterion 3.

The applicant should provide either an IP or a completed RSR. If the applicant submits an IP, it should describe the methodology for treating important human actions. The NRC will review it using the criteria in Section 7.4. Then the applicant will submit the RSR when the work described by the IP is completed.

If the applicant submits a completed RSR, the NRC will verify the results using the criteria in Section 7.4. At a minimum, the RSR should include the following:

- the final list of important HAs

- a description of the methodology employed to identify and select the HAs

Summaries may be used for any of the above items provided that references are given for more detailed documents. If the methodology was described in an IP that the NRC staff previously reviewed, the contents of the RSR should be consistent with the approved methodology and the applicant should discuss the rationale for any deviations from it.

7.4   Review Criteria

(1)   The applicant should identify risk-important HAs from the PRA/HRA.

   Additional Information: The NRC's technical branch responsible for PRA reviews the acceptability of the applicant's methodology for identifying risk-important human actions. The human factors engineering staff is responsible for ensuring that risk-important HAs included in the HFE design process are the same as those identified in Chapter 19. NRC reviewers should be aware that risk- important HAs may be distributed throughout multiple Chapter 19 tables, a practice that has caused delays in completing reviews.

(2)   Applicants should identify deterministically important HAs from the following licensing analyses:

- operator actions credited in the DCD/FSAR Chapter 15 accident and transient analyses

- operator actions identified in the D3 coping analyses performed for DCD/FSAR Chapter 7, as specified in Section 1 and 2 of Interim Staff Guidance DI&C-ISG-02, *Diversity and Defense in Depth (D3) Issues* (NRC, 2009)

Additional Information: The HFE reviewer should coordinate with the appropriate NRC technical staff to ensure that the operator actions credited in the Chapter 15 accident and transient analysis and D3 coping analyses are correctly identified.

(3) The applicant should specify how important HAs are addressed by the HFE program, in Function Allocation, Task Analysis, HSI design, Procedural Development, and Training Program Development, in order to minimize the likelihood of human error and facilitate error-detection and recovery capability.

Additional Information: The applicant's treatment of important HAs will help ensure that the design supports these actions, and that they are within acceptable human performance capabilities (e.g., within time and workload requirements).

(4) *Additional Considerations for Reviewing the HFE Aspects of Plant Modifications* - In addition to any of the criteria above that relate to the modification being reviewed, the applicant should address the following considerations:

- Whether the modification changes any of the important HAs. If so, the necessary analyses of this element should be performed to ensure that the design of the plant still addresses them appropriately.
- Whether there are new important HAs based on the modification. If so, they should be considered using the methods of this element.

Additional Information: NRC IN 97-78 and NUREG-1764 contain valuable information and guidance on modifications and important human actions.

7.5 Bibliography

DI&C-ISG-02, Rev. 2: *Digital Instrumentation and Controls, Task Working Group #2: Diversity and Defense in Depth Issues, Interim Staff Guidance* (NRC, 2009).

IEEE Std. 1082-1997: *IEEE Guide for Incorporating Human Action Reliability Analysis for Nuclear Power Generating Stations* (Institute of Electrical and Electronics Engineers, 1997).

NEI 00-04, Rev. 0: *10 CFR 50.59 SSC Categorization Guideline* (Nuclear Energy Institute, 2005).

NRC Information Notice 97-78, *Crediting of Operator Actions in Place of Automatic Actions and Modifications of Operator Actions, Including Response Times* (NRC, 1997).

NUREG-0800: *Standard Review Plan: Chapter 19, Use of Probabilistic Risk Assessment in Plant-Specific, Risk-Informed Decision Making: General Guidance* (NRC, 2007).

NUREG-1560: *Individual Plant Examination Program: Perspectives on Reactor Safety and Plant Performance* (NRC, 1997).

NUREG-1624, Rev. 1: *Technical Basis and Implementation Guidelines for a Technique for Human Event Analysis (ATHEANA)* (NRC, 2000).

NUREG-1764, Rev. 1: *Guidance for the Review of Changes to Human Actions* (Higgins, O'Hara, et al. 2007).

NUREG-1792: *Good Practices for Implementing Human Reliability Analysis* (NRC, 2005)

Regulatory Guide 1.174, Rev. 2: *An Approach for Using Probabilistic Risk Assessment in Risk-Informed Decisions on Plant-Specific Changes to the Licensing Basis* (NRC, 2011).

Regulatory Guide 1.177, Rev. 1: *An Approach for Plant-Specific, Risk-Informed Decision making: Technical Specifications* (NRC, 2011).

Regulatory Guide 1.201, Rev. 1: *Guidelines for Categorizing Structures, Systems, and Components in Nuclear Power Plants According to Their Safety Significance* (NRC, 2006).

Risk-informed inspection notebooks for each plant.

8　HUMAN-SYSTEM INTERFACE DESIGN

8.1　Background

The HSI design process represents the translation of function and task requirements into HSI characteristics and functions. The HSIs with which personnel interact should be designed through a structured methodology guiding designers in identifying and selecting candidate HSI approaches, defining the detailed design, and performing HSI tests and evaluations. The methodology should cover the development and use of HFE guidelines tailored to the unique aspects of the applicant's design, including a style guide to define the design-specific conventions (e.g., colors, symbols) that will be used in the HSI design. The availability of such an HSI design methodology will help reviewers to verify standardization and consistency in applying HFE principles. The process and the rationale for the HSI design should be documented for review.

Issues related to the detailed design of specific aspects of the HSIs should be resolved during work on the HSI design, rather than at verification and validation (V&V), at which point making modifications to the design is much more difficult. For example, decisions about acceptable display formats or alarm system processing should occur because of HSI tests and evaluations conducted during the HSI design phase rather than deferring these decisions to V&V (as described in Section 11).

8.2　Objective

The NRC staff uses the review criteria in this section to verify that the applicant has a process to translate the functional- and task-requirements to HSI design requirements, and to the detailed design of alarms, displays, controls, and other aspects of the HSI which is based on systematically applying HFE principles and criteria.

8.3　Applicant Products and Submittals

The product of the applicant's HSI design is a complete suite of HSIs that personnel use to safely operate and maintain the plant.

The applicant should provide either an IP or a completed RSR. If the applicant submits an IP, it should include a completed HSI style guide and a description of the methodology for designing the HSIs. The NRC will review this information using the criteria set out in Section 8.4 below. Then the applicant will submit the RSR when the work described by the IP is completed.

If the applicant submits a completed RSR, the NRC will verify the results using the criteria set out in Section 8.4 below. At a minimum, the RSR should include the following:

- descriptions of the inputs to the HSI design process

- the concept of how HSIs are used and an overview of HSI design

- the guidance used for the detailed HSI design

- the detailed HSI description of the main control room, technical support center, emergency operations facility, remote shutdown facility, and local control stations, covering their form, function, and performance characteristics

- a description of how the design minimizes the effects of degraded I&C and HSI conditions on the performance of personnel

- the outcomes of tests and evaluations undertaken to support the HSI design

Summaries may be used for any of the above items provided that references are given for more detailed documents. If the methodology was described in an IP that the NRC staff previously reviewed, the contents of the RSR should be consistent with the approved methodology and the applicant should discuss the rationale for any deviations from it.

8.4   Review Criteria

8.4.1   HSI Design Inputs

(1)   *Analysis of Personnel Task Requirements* – The applicant should use the following analyses, performed in earlier stages of the design process, to identify requirements for the HSIs:

- *Operational Experience Review* – An input to the HSI design should encompass lessons learned from other complex human-machine systems, especially predecessor designs and those involving similar HSI technology.

- *Functional Requirements Analysis and Function Allocation* – The HSIs should support the roles of personnel in the plant, e.g., appropriate levels of automation.

- *Task Analysis* – The set of requirements to support the role of personnel is provided by task analyses that should identify:
  - tasks needed to control the plant during a range of operating conditions from normal through accident conditions

  - detailed information and control requirements (e.g., requirements for display range, precision, accuracy, and units of measurement)

  - task support requirements (e.g., special lighting and ventilation requirements)

  - important HAs, as defined in Section 7 of this document, that should be given special attention in the HSI design process

- *Staffing and Qualifications* – The findings from analyses of staffing/qualifications should provide input for deciding upon the layout of the overall control room and allocating controls and displays to individual consoles, panels, and workstations. The staffing/qualifications analyses establish the basis for the minimum and maximum number of personnel to be accommodated, and requirements for coordinating activities between them.

(2)   *System Requirements* – The applicant should identify any constraints on the HSI design imposed by the overall I&C system, e.g., constraints on the information that can be presented due to sensor data availability.

(3)   *Regulatory Requirements* – The applicant should identify the applicable regulatory requirements as inputs to the HSI design process.

(4)   *Other Requirements* – The applicant should identify any other requirements, such as customer requirements, that are inputs to the HSI design.

8.4.2 Concept of Use and HSI Design Overview

(1)   The applicant should develop a concept of use stating the roles and responsibilities of operations personnel based upon anticipated staffing levels. The concept of use should:

   - provide a high-level description of how personnel will work with HSI resources

   - address the coordination of personnel activities, such as interactions with auxiliary operators and the coordination of maintenance and operations

   Additional Information: Examples of the types of information the applicant may identify include the allocation of tasks between the main control room or to local control stations, whether personnel will work at a single large workstation or at individual ones, to what types of information each crew member will have access, and what types of information will be displayed to the entire crew.

(2)   The applicant should provide an overview of the HSI, covering the technical bases demonstrating that they constitute a state-of-the-art HSI design supporting personnel performance. These bases may include analyses of operating experience and the literature, tradeoff studies simultaneously considering multiple alternatives, and engineering tests and evaluations. The overview should include a description of:

   - facility layouts, including workstations, large screen displays, and the nominal staff working positions

   - key HSI resources and their functionality, such as alarms, displays, controls, computer-based procedures, and other support and job aids

   - technologies to support teamwork and communication within the main control room and between the main control room, the remote shutdown facility, the TSC, EOF, and local control stations

   - the responsibilities of the crew for monitoring, interacting, and overriding automatic systems and for interacting with computerized procedures systems and other computerized operator support systems

8.4.3 HFE Design Guidance for HSIs

Applicants should employ design-specific HFE design guidance in designing the features of the HSIs, their layout, and environments. Although design guidance documents are called by different names, NUREG-0711 refers to them as "style guides." Applicants may use one or more individual documents to serve this purpose. The HFE guidelines in NUREG-0700 may serve to support the NRC staff's review of the guidance in an applicant's style guide.

(1)   The topics in the applicant's style guide(s) should address the scope of HSIs included in the design, and address their form, function, and operation, as well as the environmental conditions in which they will be used that are relevant to human performance.

   Additional Information: NUREG-0700 lists HSI topics around which a style guide(s) may be organized.

(2)   The guidance in the applicant's style guide(s) should be developed from generic HFE guidance and HSI design-related analyses. It should be tailored to reflect the applicant's design decisions in addressing specific goals of the HSI design.

        Additional Information: Analyses related to the HSI design might include an evaluation of recent literature, analysis of current industry practices and operational experience, tradeoff studies, and the findings from design-engineering experiments and evaluations.

(3)   The individual guidelines in the applicant's style guide(s) should be expressed precisely and describe easily observable HSI characteristics, such as "Priority 1 alarms are shown in red." The guidelines in the style guide(s) should be sufficiently detailed so that design personnel can deliver a consistent, verifiable design meeting the applicant's guidelines.

(4)   The applicant's style guide(s) should contain procedures for determining where and how HFE guidance will be used in the overall design process. They should be written so designers can readily understand them; the text should be supplemented with graphical examples, figures, and tables to facilitate comprehension.

(5)   The applicant should maintain the style guide(s) in a form that is readily accessible and usable by designers, and is easily modified and updated as the design matures. The guidance should include a reference(s) to the source upon which it is based.

8.4.4   HSI Detailed Design and Integration

The criteria in this section are divided into the following subsections:

        8.4.4.1, General
        8.4.4.2, Main Control Room
        8.4.4.3, Technical Support Center
        8.4.4.4, Emergency Operations Facility
        8.4.4.5, Remote Shutdown Facility
        8.4.4.6, Local Control Stations

Many criteria in this section are based on HFE guidance from other documents. We listed these documents and give the full references for them, including the specific revision or year of publication, in Section 14, References.

8.4.4.1   General

(1)   For important HAs (see Element 7), the applicant's design should minimize the probability that errors will occur, and maximize the probability that any error made will be detected.

(2)   The applicant should base the layout of HSIs within consoles, panels, and workstations on (1) analyses of personnel roles (job analysis), and (2) systematic strategies for organization, such as arrangement by importance, and frequency and sequence of use.

(3)   The applicant should design the HSIs to support inspection, maintenance, test, and repair of (1) plant equipment, and (2) the HSIs. The applicant should design the latter so that inspection, maintenance, test, and repair of the HSIs do not interfere with other

plant-control activities (e.g., maintenance tags should not block the operators' views of plant indications).

(4) The applicant's design should support personnel task performance under conditions of minimum-, typical-, and high-level or maximum staffing.

Additional Information: Minimum staffing is that defined by plant's technical specifications. Typical staffing is that specified and used by the licensee for routine plant operations. Maximum staffing includes the augmented staff for accident situations.

(5) The applicant's design process should account for using the HSIs over the duration of a shift where decrements in human performance due to fatigue may be a concern.

Additional Information: As an example, simulation tests can evaluate fatigue caused by using touch screens for long periods.

(6) The characteristics of the applicant's HSIs should support human performance under the full range of environmental conditions, ranging from normal to credible extreme conditions, such as loss of lighting and of ventilation. For the remote shutdown facility and local control stations, the applicant's HFE design should consider the ambient environment (e.g., noise, temperature, contamination) and the need for and type of protective clothing.

Additional Information: For example, consideration should be given to the effects that protective clothing may have on task performance (e.g., protective gloves may make manual dexterity tasks more difficult and increase the time necessary to complete them).

(7) The applicant should identify how in an operating plant:

- the HSIs are modified and updated

- temporary HSI changes are made (such as modifying the set points)

- personnel-defined HSIs are created (such as temporary displays that personnel define for monitoring a specific situation)

(8) *Additional Considerations for Reviewing the HFE Aspects of Plant Modifications* – In addition to any of the criteria above that relate to the modification being reviewed, the applicant should address the following considerations:

- Modified and new HSIs should be designed consistently with the style guide used for existing ones, so that personnel have a similar interface across new and old equipment.

- HSI modifications should be designed, as far as possible, to be consistent with users' existing strategies for gathering and processing information and executing actions identified in the task analysis.

  Additional Information: Consistency with existing strategies can reduce the learning personnel need to become proficient in using the modification, and therefore, the potential for errors.

- If there are changes in the degree of integration between plant systems, then the applicant should verify that the HSIs support personnel in controlling these altered

53

systems. The design of the HSIs should reasonably assure that the relationships between plant systems are depicted clearly and accurately.

8.4.4.2   Main Control Room

In some of the criteria below, we italicize and underline the word "how" to emphasize it. The word refers to the means by which the information identified in the criterion is displayed by the HSIs to personnel, e.g., how displays depict the information that operators need for monitoring tasks.

(1)   *Safety Parameter Display System* – The applicant should describe the safety parameter display system (SPDS), addressing the following:

- *Identification of Critical Safety Functions (CSFs)* – The CSFs needed to meet the requirement for an SPDS should be identified. NUREG-1342 Section III.F, Minimum Parameters for Display, lists the five CSFs that personnel monitor using an SPDS for boiling water reactor (BWRs) and pressurized water reactor (PWRs). For new designs, applicants should verify that these CSFs are suitable for their design, identifying any changes needed based on their design's detailed characteristics. CSFs may differ for non-light water reactor designs, such as high-temperature gas-cooled reactors and liquid-metal reactors.

- *Identification of the Parameters Personnel will use to Monitor Each CSF* – The applicant should identify the plant parameters personnel need to monitor each CSF and describe the means by which plant data are synthesized, combined, or otherwise evaluated to provide the information presented in the SPDS display. Section III.F of NUREG-1342 has guidance on acceptable parameters for the current fleet of PWRs and BWRs. The applicant's identification of parameters should consider the unique characteristics of the plant's design.

- *Evaluation of SPDS HSIs* – The applicant should verify that the SPDS HSIs conform to acceptable HFE practices using NUREG-0700, Section 5 and other SPDS HFE guidance.

Additional Information: SPDS requirements are described in 10 CFR 50.34(f)(2)(iv), and related guidance in NUREG-0835, NUREG-1342, Supplement 1 of NUREG-0737, and NUREG-0700, Section 5. These NUREGs discuss the NRC's review guidance for SPDS, with NUREG-0700 being the primary one; the others encompass supplemental guidance, examples, and technical bases.

(2)   *Bypassed and Inoperable Status Indication* – The applicant should describe *how* the HSI assures the automatic indication of the bypassed and inoperable status of a safety function, and the systems actuated or controlled by the safety function. [10 CFR 50.34(f)(2)(v) - I.D.3] Regulatory Guide 1.47 includes the following guidance related to the display of bypassed and inoperable status of safety systems:

- The status indication should be in the main control room.

- Administrative procedures should be supplemented by an automatic indication system that shows, for each affected safety system or subsystem, the bypass or deliberately induced inoperability of a safety function, and the systems it actuates or controls.

- Provisions should be made allowing the operations staff to confirm that a bypassed safety function was properly returned to service.

- Annunciating functions for system failure and automatic actions based on the self-test or self-diagnostic capabilities of digital computer-based I&C safety systems should be consistent with the above bullets.

- The indication system for bypass and inoperable status should include the ability to ensure its operable status during normal plant operation to the extent to which the indicating and annunciating functions can be verified.

- Bypass and inoperable status indicators should be arranged such that personnel can determine whether it is permissible to continue operating the reactor.

- The control room of all affected units should receive an indication of the bypass for their shared system safety functions.

(3)   *Relief and Safety Valve Position Monitoring* – The applicant should describe *how* the HSI indicates the position of the relief and safety valves (open or closed) in the control room. [10 CFR 50.34(f)(2)(xi)- II.D.3]

(4)   *Manual Feedwater Control* – The applicant should describe *how* the HSI provides automatic and manual initiation of the auxiliary feedwater system, and indicates auxiliary feedwater system flow in the control room. [Applicable to PWRs only, 10 CFR 50.34(f)(2)(xii) - II.E.1.2]

(5)   *Containment Monitoring* – The applicant should describe *how* the control room's HSIs (alarms and displays) inform personnel about: (A) containment pressure; (B) containment water level; (C) containment hydrogen concentration; (D) containment radiation intensity (high level); and (E) noble gas effluents for all potential, accident release points. [10 CFR 50.34(f)(2)(xvii) - II.F.1]

(6)   *Core Cooling* – The applicant should describe *how* the HSI provides unambiguous indication of inadequate core cooling, such as with primary coolant saturation meters in PWR's, and a suitable combination of signals from indicators of coolant level in the reactor vessel and in-core thermocouples in PWRs and BWRs. [10 CFR 50.34(f)(2)(xviii) - II.F.2]

(7)   *Post-accident Monitoring* – The applicant should describe *how* the HSI assures monitoring of plant and environmental conditions following an accident including core damage. [10 CFR 50.34(f)(2)(xix) - II.F.3, and RG1.97]

(8)   *Auxiliary Heat Removal* -- The applicant should describe *how* that necessary automatic and manual actions can be taken to ensure proper functioning of auxiliary heat removal systems when the main feedwater system is not operable. [Applicable to BWRs only, 10 CFR 50.34(f)(2)(xxi) - II.K.1.22]

(9)   *Reactor Level Monitoring* – The applicant should describe *how* the HSI gives a record of the reactor vessel's water level in one location on displays that meet normal post-accident recording requirements. [Applicable to BWRs only, 10 CFR 50.34(f)(2)(xxiv) - II.K.3.23]

(10) *Leakage Control* – The applicant should describe *how* the HSI provides for leakage control and detection in the design of systems outside containment that contain (or might contain) accident-source-term radioactive materials after an accident. [10 CFR 50.34(f)(2)(xxvi) - III.D.1.1]

(11) *Radiation Monitoring* – The applicant should describe *how* the HSI provides appropriate monitoring of in-plant radiation and airborne radioactivity under a broad range of routine- and accident conditions. [10 CFR 50.34(f)(2)(xxvii) - III.D.3.3]

(12) *Manual Initiation of Protective Actions* – The applicant should describe *how* the HSI supports the manual initiation of protective actions at the system level for safety systems otherwise initiated automatically. [Regulatory Guide 1.62.]

(13) *Diversity and Defense-in-depth* – The applicant should describe *how* the HSI provides displays and controls in the MCR for manual, system-level actuation of critical safety functions, and for monitoring those parameters that support them. These displays and controls are independent of, and different from, the normal I&C. [I&C BTP7-19, Point 4]

(14) *Important HAs* – The applicant should describe *how* the HSI provides the controls, displays, and alarms that ensure the reliable performance of identified important HAs. Section 7 of this document discusses important HAs.

(15) *Computer-Based procedure platform* - The applicant's computer-based procedures should be consistent with the design review guidance in NUREG-0700, Section 8, Computer-Based Procedure System and in Section 1 of DI&C-ISG-5 (NRC, 2008).

8.4.4.3 Technical Support Center

NUREG-0696 states that HFE should be incorporated in the design of the on-site Technical Support Center (TSC), and considers both operating and maintenance personnel. The criteria in this section are applicable to the HFE aspects of the review of the TSC. The applicant's submittal should include the following:

(1) The applicant should describe *how* the HSIs give personnel the information needed to:

- analyze the plant's steady-state and dynamic behavior before and throughout an accident so TSC personnel can guide the MCR operators in managing the abnormal conditions and mitigating the accident without interfering with the MCR activities
- undertake the needed environmental- and radiological-monitoring functions of the EOF when it is not operational
- offer technical support to personnel during recovery operations after an emergency
- provide reliable voice-communications facilities to the control room, the operations support center, the EOF, the NRC, and with state and local operations centers

(2) The applicant should describe *how* the HSIs give personnel the information needed for:

- determining the plant's steady-state operating conditions before the accident
- ascertaining the transient conditions producing the initiating event
- gauging plant systems' dynamic behavior throughout the accident

- reviewing the accident sequence
- deciding upon appropriate mitigating actions
- evaluating the extent of any damage
- assessing the plant's status during recovery operations

(3) The applicant should describe *how* the HSIs provide an SPDS that replicates the SPDS in the MCR (to improve the exchange of information between personnel in the main control room and the EOF). If the SPDS in the main control room is composed of multiple displays, then multiple displays also should be provided in the TSC.

(4) The applicant should describe *how* the HSIs provide as a minimum, the set of variables specified in Regulatory Guide 1.97, Revision 4, plus all sensor data and calculated variables not specified in Reg. Guide 1.97 but included in the data sets for the SPDS, for the EOF, or for transmission to offsite locations.

(5) The applicant should describe *how* the HSIs allow all TSC personnel to complete their assigned tasks with unhindered access to alphanumeric and/or graphical representations of:

- plant systems variables
- in-plant radiological variables
- meteorological information
- offsite radiological information

(6) The applicant should describe *how* the HSIs provide the trend-information displays and time-history displays that give the TSC personnel a dynamic view of the plant's status during abnormal operating conditions.

(7) The applicant should describe *how* HFE was incorporated into the TSC design to ensure that personnel easily understand and use the HSIs.

8.4.4.4 *Emergency Operations Facility*

NUREG-0696 states that HFE should be incorporated in the design of the Emergency Operations Facility (EOF) considering both operating and maintenance personnel. The criteria in this section are applicable to the HFE review of the EOF.

(1) The applicant should describe *how* the HSIs assure the acquisition, display, and evaluation of all radiological, meteorological, and plant-system data essential to determining offsite protective measures.

(2) The applicant should describe *how* the HSIs continuously indicate radiation dose-rates and concentrations of airborne radioactivity inside the EOF while it is used during an emergency, including local alarms with trip levels set to provide early warning to EOF personnel of adverse conditions that may affect the facility's habitability.

(3)   The applicant should describe *how* the HSIs support reliable voice communications to the TSC, the main control room, the NRC, and the state and local emergency response facilities.

(4)   The applicant should describe *how* the HSIs supply data sufficient to assess the actual and potential onsite and offsite environmental consequences of an emergency, and information on the general condition of the plant.

(5)   The applicant should describe *how* the HSIs provide radiological, meteorological, and other environmental data to:

- assess environmental conditions

- coordinate radiological monitoring

- recommend implementing offsite emergency plans

As a minimum, the EOF data should include (1) sensor data of the variables specified in Reg. Guide 1.97, Revision 4, and (2) the meteorological variables specified in the proposed Revision 1 to Regulatory Guide 1.23, "Meteorological Measurements Programs in Support of Nuclear Power Plants," and in NUREG-0654, Revision 1, Appendix 2.

(6)   The applicant should describe *how* the EOF HSIs provide all data that are available for display in the TSC, including information sent from the plant to the NRC.

(7)   The applicant should describe *how* the HSIs allow all EOF personnel to perform their assigned tasks with unhindered access to alphanumeric and/or graphical representations of:

- plant system variables

- in-plant radiological variables

- meteorological information

- offsite radiological information

(8)   The applicant should describe *how* the HSIs display the needed trend information and time-history data in the EOF. The displays should be partitioned to facilitate the different functional groups in the EOF retrieving this information.

(9)   The applicant should describe *how* the HSIs provide an SPDS to improve the exchange of information between the MCR and the TSC. If the SPDS in the MCR comprises multiple displays, they should also be provided in the EOF.

(10)   The applicant should describe *how* HFE was incorporated into the EOF design to ensure that personnel easily understand and use the HSIs.

8.4.4.5 Remote Shutdown Facility

(1) The applicant should describe *how* the HSI provides a design capability for remote shutdown of the reactor outside the main control room. [10 CFR 50, Appendix A, General Design Criteria 19]

(2) The applicant should describe *how* the HSIs at the remote shutdown facility are consistent with those in the main control room.

8.4.4.6 Local Control Stations

(1) The applicant should describe the basis for deciding which HSIs will be included in the main control room design, and which will be provided locally.

(2) The applicant should describe *how* HFE was incorporated into the HSIs for local control stations to ensure they are consistent with those in the MCR, and that personnel easily understand and use the HSIs.

8.4.5 Degraded I&C and HSI Conditions

(1) The applicant should identify:

- the effects of automation failures and degraded conditions on personnel and plant the performance

- HFE-significant I&C degradations; i.e., the failure modes and degraded conditions of the I&C system that might adversely affect the HSIs personnel use to accomplish important HAs

   Additional Information: The I&C system is made up of four subsystems: Sensor, monitoring, automation and control, and communications. In this criterion, automation is considered separately due to its well-known human performance challenges and their potential impact on safety. The focus of this criterion is on HFE-significant I&C degradations. An example is a sensor degradation that results in a control room display that confuses personnel into thinking there is a process disturbance.

(2) The applicant should specify the alarms and other information personnel need to detect degraded I&C and HSI conditions in a timely manner, and to identify their extent and significance.

(3) The applicant should determine any needed back-up systems to ensure that important personnel tasks can be completed under degraded I&C and HSI conditions.

(4) The applicant should determine the necessary compensatory actions and supporting procedures to ensure that personnel effectively manage degraded I&C and HSI conditions, and the transition to back-up systems.

8.4.6 HSI Tests and Evaluations

Tests and evaluations (T&Es) of concepts and detailed design features are conducted during the process of developing HSIs to support design decisions. This section provides review guidance for two types of T&Es:

- Trade-off evaluations are comparisons between design options, based on aspects of human performance that are important to successful task performance, and to other design considerations.

- Performance-based tests involve assessing personnel performance, including subjective opinions, to evaluate design options and design acceptability.

8.4.6.1 Trade-off Evaluations

(1) In comparing design approaches, the applicant should consider those aspects of human performance important to performing tasks. The applicant should take into account the following factors when developing criteria to apply in selecting one design approach over another:

- personnel-task requirements

- human-performance capabilities and limitations

- HSI-system performance requirements

- inspection and testing needs

- maintenance demands

- use of proven technology and the operating experience of predecessor designs

Additional Information: Including selection criteria for human performance will help to ensure that the differential effects of design options on human performance can be assessed, along with other considerations. For example, when analyzing trade-offs between using either a mouse or a touch screen as a computer-input device, the fatigue caused by using the device, and the time required to perform actions using each device should be considered.

(2) The applicant should state explicitly the relative benefits of design alternatives and the basis for the design approach selected.

8.4.6.2 Performance-Based Tests

(1) The applicant should identify the specific objectives of the tests.

Additional Information: Performance-based tests have many different purposes, such as choosing between design alternatives or verifying that an aspect of the HSI meets performance criteria.

(2) The applicant should base the general approach to testing on the test's objective(s). The following aspects of the tests should be described (note that not all items are applicable to every type of test):

- participants
- testbed
- design features or characteristics of the HSI being tested
- tasks or scenarios used
- performance measures
- test procedures
- data analyses

(3)   The conclusions from the tests and their impact on design decisions should be described.

Additional Considerations for Reviewing the HFE Aspects of Plant Modifications – The applicant should address any of the criteria above that relate to the modification being reviewed, including the following considerations:

- The extent to which HSI modifications are consistent with users' existing HSIs and the licensee's Safety Analysis Report and HFE commitments.
- The extent to which HSI modifications support crew coordination.

8.5   Bibliography

ANSI/AIAA G-035A-2000: *Guide to Human Performance Measurements* (AIAA, 2001).

ANSI/HFES 100-2007: *Human Factors Engineering of Computer Workstations* (HFES, 2007).

ANSI/HFES 200: *Human Factors Engineering of Software User Interfaces* (HFES, 2008).

BNL TR E2090-T4-1-9/96: *Human-System Interface Design Process and Review Criteria* (Stubler & O'Hara, 1996).

BNL TR E2090-T4-4-12/94, Rev. 1: *Group-View Displays* (Stubler & O'Hara, 1996).

BNL TR 91017-2010: *Human-System Interfaces to Automatic Systems: Review Guidance and Technical Basis* (O'Hara & Higgins, 2010).

DI&C-ISG-05 (Revision1): *Highly-Integrated Control Rooms—Human Factors Issues* (NRC, 2008).

IEC 60964: *Nuclear Power Plants-Control Rooms - Design* (International Electrochemical Commission, 2009).

IEEE Std. 1023-2004: *IEEE Recommended Practice for the Application of Human Factors Engineering to Systems, Equipment, and Facilities of Nuclear Power Generating Stations and Other Nuclear Facilities* (Institute of Electrical and Electronics Engineers, 2004).

NUREG-0654: *Criteria for Preparation and Evaluation of Radiological Emergency Response Plans and Preparedness in Support of Nuclear Power Plants* (NRC, 1980).

NUREG-0696: *Functional Criteria for Emergency Response Facilities* (NRC, 1981).

NUREG-0700: *Human-System Interface Design Review Guidelines* (NRC, 2002).

NUREG-0737: *Clarification of TMI Action Plan Requirements* (NRC, 1980).

NUREG-0800, Branch Technical Position 7-19, Rev. 5: *Guidance for Evaluation of Diversity and Defense-In-Depth in Digital Computer-Based Instrumentation and Control Systems* (NRC, 2007).

NUREG-0835: *Human Factors Acceptance Criteria for the Safety Parameter Display System* (NRC, 1981).

NUREG-1342: *A Status Report Regarding Industry Implementation of Safety Parameter Display System* (NRC, 1989).

NUREG/CR-6393: *Integrated System Validation: Methodology and Review Criteria* (O'Hara, et al., 1997).

NUREG/CR-6633: *Advanced Information Systems: Technical Basis and Human Factors Review Guidance* (O'Hara, Higgins & Kramer, 2000).

NUREG/CR-6634: *Computer-Based Procedure Systems: Technical Basis and Human Factors Review Guidance* (O'Hara, Higgins, Stubler & Kramer, 2000).

NUREG/CR-6635: *Soft Controls: Technical Basis and Human Factors Review Guidance* (Stubler, O'Hara & Kramer, 2000).

NUREG/CR-6636: *Maintenance of Digital Systems: Technical Basis and Human Factors Review Guidance* (Stubler, Higgins & Kramer, 2000).

NUREG/CR-6637: *Human-System Interface and Plant Modernization Process: Technical Basis and Human Factors Review Guidance* (Stubler, O'Hara, Higgins & Kramer, 2000).

NUREG/CR-6684: *Advance Alarm Systems: Guidance Development and Technical Basis* (Brown, et al., 2000).

Regulatory Guide 1.23, Rev 1: *Meteorological Measurements Programs in Support of Nuclear Power Plants* (NRC, 2007).

Regulatory Guide 1.47: *Bypassed and Inoperable Status Indication for Nuclear Power Plant Safety Systems*, Rev 1 (NRC, 2010).

Regulatory Guide 1.62: *Manual Initiation of Protective Actions* (NRC, 2010).

Regulatory Guide 1.97, Rev 4: *Criteria For Accident Monitoring Instrumentation For Nuclear Power Plants* (NRC, 2006).

UCRL-15673: *Human Factors Design Guidelines for Maintainability of Department of Energy Nuclear Facilities* (Bongarra, et al., 1985).

9   PROCEDURE DEVELOPMENT

9.1   Background

Procedures are essential to plant safety because they support and guide personnel interactions with plant systems and personnel responses to plant-related events. In the nuclear industry, procedure development is the responsibility of individual utilities. The procedures program is reviewed by NRC staff using SRP Chapter 13. The NRC is considering adding procedures to the list of operational programs in SRP Chapter 13.4.

Procedures should be supported by the analyses used to develop the HSIs and training. All three elements should be subject to a common evaluation process that verifies the three elements work together to maximize operator performance. To ensure complete integration and consistency, the same HFE principles should be applied to procedures and other HSIs provided to personnel.

For new plant designs and advanced reactors, the generic technical guidelines (GTG) and procedures should receive input from the analyses used to develop the HSIs and training to obtain a high degree of integration and consistency. For existing operating plants, the GTG for emergency operating procedures was developed by BWR- and PWR-owner groups working with the nuclear steam supply system (NSSS) vendors. NSSS vendors are likely to again play a role in creating the GTG for advanced plants. For plants that modernize, the procedural modifications should address all personnel tasks affected by the changes in plant systems and HSIs. Procedures should be developed or modified to reflect the characteristics and functions of such modifications.

9.2   Objective

The procedure element is integral to an overall HFE program and should be developed and implemented using accepted human factors engineering principles.

The objective of the NRC procedure review is to confirm that the applicant's procedure development program incorporates HFE principles and criteria, along with all other design requirements, to develop procedures that are technically accurate, comprehensive, explicit, easy to utilize, validated, and in conformance with 10 CFR 50.34(f)(2)(ii).

9.3   Applicant Products and Submittals

The products of the applicant's procedure development program are the generic technical guidelines, the procedure writers' guides, and the full set of plant procedures. This material should conform to the acceptance criteria specified in SRP, Chapter 13 and should be submitted by the applicant in accordance with the guidance in Chapter 13. No procedure related submittal is expected as part of the Chapter 18 material.

9.4   Review Criteria

These review criteria, which emphasize accepted human factors engineering principles, are provided as information. They are a subset of those provided in the SRP, Chapter 13.

(1)   The scope of the applicant's procedure development program should include:

- the GTG for emergency operating procedures (EOPs)
- plant and system operations (including startup, power, and shutdown operations)
- test and maintenance
- surveillance testing
- abnormal and emergency operations
- alarm response

(2) The applicant should identify the basis for developing procedures, which should include:

- plant-design bases
- system-based technical requirements and specifications
- results of task analyses
- important HAs
- initiating events to be considered in the EOPs, including those in the design bases
- the GTG for EOPs
- appropriate HFE of procedures

(3) The applicant should develop a writer's guide to establish the process for developing technical procedures that are complete, accurate, consistent, and easy to understand and follow. The guide should contain:

- objective criteria, so that the procedures, developed in accordance with it, are consistent in organization, style, and content
- instructions for procedure content and format, including writing of the action steps and specifying acceptable lists of abbreviations/acronyms and terms to be used

The applicant should use the guide for all procedures within the scope of this element.

(4) The applicant's procedures should contain the following elements, as applicable:

- title and identifying information, such as number, revision, and date
- statement of applicability and purpose
- prerequisites
- precautions (including warnings, cautions, and notes)
- important human actions
- limitations and actions
- acceptance criteria
- check off lists
- reference material

(5)   The applicants should develop symptom-based GTG and EOPs with clearly specified entry conditions.

(6)   The applicant should:

- Verify that the procedures are correct, and can be carried out by plant personnel. Wherever possible, this should include a walkdown of the procedure either on a full-scope simulator, or in the facility itself. Where a walkdown is not possible, a tabletop verification may be used.

- Validate the use of procedures in a simulation of the integrated system as part of the activities described in Section 11, Human Factors Verification and Validation.

- When procedures are modified, the applicant should verify the adequacy of their content, format, and integration. The applicant also should validate procedures when a modification substantially changes personnel tasks significant to plant safety. The validation should assure that the procedures correctly reflect the characteristics of the modified plant, and can be carried out effectively to operate or maintain the plant.

(7)   The applicant's computer-based procedures should be consistent with the design review guidance in NUREG-0700, Section 8, Computer-based Procedure System and in Section 1 of DI&C-ISG-5 (NRC, 2008).

(8)   The applicant should have a plan for maintaining procedures and controlling updates. Procedure modifications should be integrated across the full set of procedures. Changes in particular parts of the procedures should not conflict with other parts nor be inconsistent with them.

(9)   The applicant should evaluate the physical means by which personnel access and use procedures, especially during operational events.

Additional Information: This criterion generally applies to both hard-copy and computer-based procedures, although the nature of the issues differs somewhat depending on their implementation. For example, the applicant should address the storage of procedures, ease of the operator's access to the correct procedures, and laydown of hard-copy procedures for use in the MCR, the remote shutdown facility, and local control stations.

9.5   Bibliography

ANS 3.2-2006: *Administrative Controls and Quality Assurance for the Operational Phase of NPPs* (American Nuclear Society, 1994).

DI&C-ISG-05 (Revision1): *Highly-Integrated Control Rooms—Human Factors Issues* (NRC, 2008).IP 42700: *Plant Procedures*. (NRC, periodically updated).

IP 42001: *Emergency Operating Procedures*. (NRC, periodically updated).

NRC Regulatory Guide 1.33 (Rev. 2): *Quality Assurance Program Requirements* (NRC, 1978).

NUREG/CR-6634: *Computer-Based Procedure Systems: Technical Basis and Human Factors Review Guidance* (O'Hara, Higgins, Stubler & Kramer, 2000).

NUREG/CR-5228: *Techniques for Preparing Flowchart Format Emergency Operating Procedures*, Volumes 1 and 2 (Barnes et al., 1989).

NUREG-0800: *Standard Review Plan* (NRC, 2007).

NUREG-0899: *Guidelines for the Preparation of Emergency Operating Procedures* (NRC, 1982).

NUREG-1358: *Lessons Learned From the Special Inspection Program for Emergency Operating Procedures* (NRC, 1989), and Supplement 1 (NRC, 1992).

Procedure Writing: Principles and Practices (Wieringa, Moore & Barnes, 1998).

10.4.2 Organization of Training

(1)   The applicant should define the roles of all organizations for developing the training requirements, the training information sources, and the materials for training, and thereafter, implementing the training program.

Additional Information: For example, the role of a vendor may range from merely providing input materials (e.g., the GTG), to conducting portions of specific training programs.

(2)   The applicant should define the qualifications of organizations and personnel involved in developing and conducting training.

(3)   The applicant should define the facilities and resources needed to satisfy the requirements of the training program, and the guidance in ANSI 3.5(ANS, 2009) and Regulatory Guide 1.149 (NRC, 2011), such as plant-referenced, full-scope and part-task training simulators.

10.4.3 Learning Objectives

(1)   The applicant should derive learning objectives from the analysis describing the desired performance after training. This analysis should include, but not be limited to, the training needs identified in the following:

- *Licensing Basis* – Final Safety Analysis Report, system description manuals and operating procedures, facility license and license amendments, licensee event reports, and other documents identified by the NRC staff as being important to training

- *Operating Experience Review* – previous training deficiencies and operational problems that may be corrected through additional or enhanced training, and the positive characteristics of previous training programs

- *Functional Requirements Analysis and Function Allocation* – functions identified as new or modified

- *Task Analysis* - tasks identified during task analysis as posing unusual demands, including new or different tasks, and tasks requiring a high degree of coordination, high workload, or special skills

- *Treatment of Important Human Actions* – coordinating individual roles to reduce the likelihood and/or consequences of human error associated with important HAs, and the use of advanced technology

- *HSI Design* – design features whose purpose or operation may differ from the past experience or expectations of personnel

- *Plant Procedures* – tasks that were identified as problematic in developing procedures (e.g., procedural steps that underwent extensive revision resulting from concerns about plant safety)

- *Verification and Validation* (V&V) – training concerns identified during V&V, including HSI usability issues noted during validation or verification, and problems with personnel performance identified during validation trials (e.g., misdiagnoses of plant events)

(2)   The applicant's learning objectives for personnel training should address the knowledge and skill needs and attributes of all relevant dimensions of the trainee's job, such as interactions with the plant, the HSIs, and other personnel.

Additional Information: Table 10-1 illustrates these dimensions.

Table 10-1 Some Knowledge and Skill Dimensions for Learning Objectives Identification

Topic	Knowledge	Skill
Plant Interactions	Understanding of plant processes, systems, operational constraints, and failure modes	Skills associated with monitoring and detection, situation awareness, response planning, and implementation
HSI and Procedure Interactions	Understanding of procedures and HSI structure, functions, failure modes, and interface-management tasks (actions, errors, and recovery strategies)	Skills associated with interface-management tasks
Personnel Interactions (In the MCR and in the plant)	Understanding information requirements of others, how actions should be coordinated with others, policies and constraints on personnel interactions	Skills associated with personnel interactions (i.e., teamwork)

10.4.4 Design of the Training Program

(1)   The applicant should define how learning objectives will be conveyed to the trainee. The definition should include:

- the use of lectures, simulators, and on-the-job training to convey particular categories of learning objectives

- specific plant conditions and scenarios to be used in training programs

- training implementation, such as the temporal order and schedule of the training segments

(2)   The applicant's training of reactor operators using nuclear power plant simulation facilities should conform to Regulatory Guide 1.149. The applicant should provide the details of the program for simulator training, including length of time (weeks), and a description of the simulation facility as required by 10 CFR 55.45(b) and 55.46.

10.4.5 Content of Training Program

(1)   The applicant's training of factual knowledge should be taught using actual tasks so personnel learn to apply it in the work environment. The context of the job should be defined, and represented realistically to help trainees link this knowledge to the job's requirements. Training addressing theory should be integrated with training in using procedures.

(2) The applicant's training of skills should be structured so that the environment is consistent with the level of skill being taught. It should support the acquisition of skills by allowing trainees to manage cognitive demands.

Additional Information: For example, trainees should not be placed in environments teaching high-level skills, such as coordinating control actions among crew members, before they have mastered requisite, low-level skills, such as how to manipulate control devices.

(3) The applicant's training should address rules for decision-making for plant systems, HSIs, and procedures. It should include rules for accessing and interpreting information, and for interpreting the symptoms of failures of systems, HSIs, and procedures. This training should cover acquiring new decision-making rules, and eliminating existing ones that are inappropriate to the design.

(4) The applicant's training for performance under degraded conditions should support personnel in:

- understanding how and why the I&C subsystems might degrade or fail

- knowing the implications of degradations in the HSIs for their own task performance

- monitoring the I&C system's performance, so degradations are detected and recognized via the control room's HSIs

- performing recovery actions and compensatory actions in the event of a degraded condition, for example through the use of procedures

- smoothly transitioning to backup systems when needed

- comprehending how the roles and responsibilities of personnel and the concept of use will be impacted

10.4.6 Evaluation and Modification of Training

(1) The applicant should define:

- the methods for evaluating the overall effectiveness of the training programs and trainee mastery of training objectives, including written- and oral-tests and reviews of performance during walkthroughs, simulator exercises, and job performance

- evaluation criteria for training objectives for individual training modules

- methods for assessing overall proficiency and coordination with any applicable regulations

(2) The applicant should define the methods for verifying the accuracy and completeness of the training course materials.

(3) The applicant should establish procedures for refining and updating the content and conduct of training, including procedures for tracking modifications in the training courses.

10.4.7 Periodic Retraining

(1)　The applicant's program should contain provisions to periodically retrain personnel.

(2)　The applicant should evaluate whether any changes in retraining are warranted following plant modernization programs.

10.5　Bibliography

10 CFR Part 55: *U.S. Code of Federal Regulations*, Part 55, "Operators' Licenses," Title 10, "Energy."

10 CFR 50.120: *U.S. Code of Federal Regulations*, Part 50, "Training and Qualification of Nuclear Power Plant Personnel."

ANSI/ANS 3.1-1993; R1999: *Selection, Qualification, and Training of Personnel for Nuclear Power Plants* (American Nuclear Society, 1999).

ANSI/ANS 3.5-2009: *Nuclear Power Plant Simulators for Use in Operator Training* (American Nuclear Society, 2009).

IP 41500: *Training and Qualification Effectiveness.* (NRC, periodically updated).

NUREG-0800: *Standard Review Plan*, Section 13.2, Training, (NRC, 2007).

NUREG-1021: *Operator Licensing Examination Standards for Power Reactors*, Rev. 9, (NRC, 2004) and Supplement 1, 2007 (NRC, 2007).

NUREG-1220: *Training Review Criteria and Procedures* (NRC, Revised periodically).

Regulatory Guide 1.149 (Rev 4): *Nuclear Power Plant Simulation Facilities for Use in Operator Training and License Examinations* (NRC, 2011).

Regulatory Guide 1.8: *Personnel Selection and Training* (NRC, 2000).

11   HUMAN FACTORS VERIFICATION AND VALIDATION

11.1   Background

V&V evaluations comprehensively determine that the HFE design conforms to HFE design principles and that it enables plant personnel to successfully perform their tasks to assure plant safety and operational goals. The V&V element consists of four major activities: Sampling of Operational Conditions, Design Verification, Integrated System Validation, and HED Resolution (Figure 11-1).

Sampling of Operational Conditions to support V&V tests is important because reviews of new plants and significant HSI modifications can involve hundreds or thousands of individual HSIs, and it is impractical and unnecessary to review all of them. Therefore, the applicant can employ a sampling strategy to guide the selection of HSIs to review.

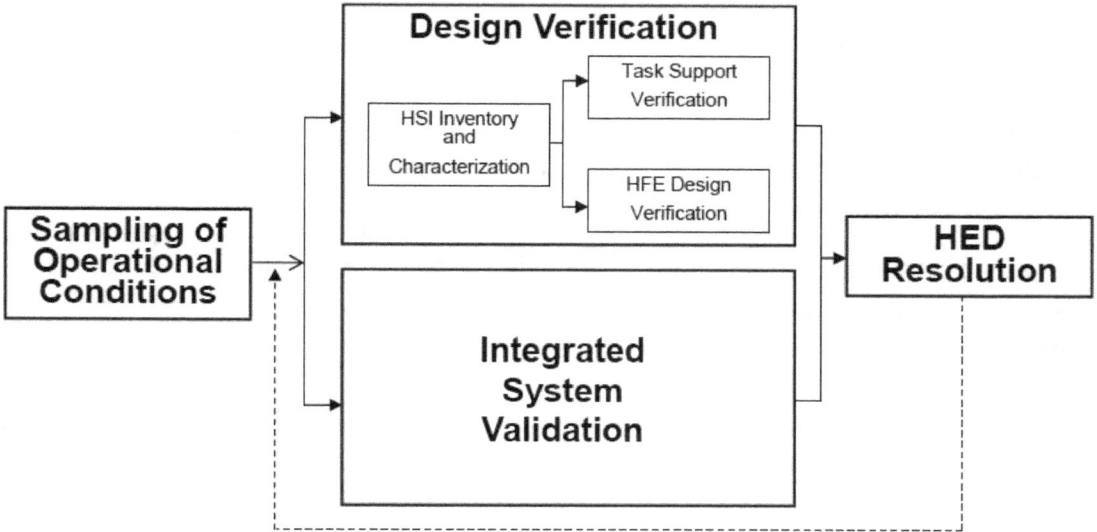

Figure 11-1 Overview of verification and validation activities

The review involves two types of Design Verification: HSI Task Support Verification and HFE Design Verification. The former is an evaluation to verify that the HSI supports the requirements of personnel tasks, as defined by task analyses. HEDs are identified for (1) personnel task requirements that the HSIs do not fully support, and (2) the presence of HSIs that may not be needed to support personnel tasks. The latter, HFE Design Verification, is an evaluation to verify that the HSIs are designed to accommodate human capabilities and limitations as reflected in HFE guidelines, such as those in NUREG-0700. HEDs are identified if the design is inconsistent with them.

Integrated System Validation (ISV) is an evaluation, using performance-based tests, to determine whether an integrated system's design (i.e., hardware, software, and personnel elements) meets performance requirements and supports the plant's safe operation.   HEDs are identified if performance criteria are not met.

HED Resolution is an evaluation to provide reasonable assurance that the HEDs identified during the V&V have been assessed and resolved. HED Resolution should be performed iteratively with V&V; that is, the applicant may address and resolve issues identified during one V&V activity before conducting other V&V activities. The preferred order is HSI Task Support Verification, HFE Design Verification, and ISV, although iteration may be needed.

Many design documents (e.g., ISO 11064) recommend conducting V&V throughout the design process. NUREG-0711 agrees with that recommendation, but with these activities being called "HSI Tests and Evaluations" (see the HSI Design element, Section 8.4.6). As such, they are distinguishable from V&V since they are activities whereby issues on HSI subsystem design (such as the coding techniques used in the alarm system) are explored and evaluated. V&V, as used in NUREG-0711, is considered a test that final design requirements are met.

There are separate NRC reviews to validate procedures and training programs conducted as part of Chapter 13 of the Standard Review Plan.

11.2 Objectives

The NRC staff uses the review criteria in this section to verify that:

- The applicant identified a sample of operational conditions that (1) includes conditions representative of the range of events that could be encountered during the plant's operation, (2) reflects the characteristics expected to contribute to variations in the system's performance, and (3) considers the safety significance of HSIs. These sample characteristics are best identified by using a multidimensional sampling strategy to reasonably assure that V&V evaluations include variation along important dimensions.

- The applicant's HSI inventory and characterization accurately describes all HSI displays, controls, and related equipment lying within the scope defined by the sampling of operational conditions.

- The applicant verified that the HSI provides the needed alarms, information, controls, and task support defined by task analysis for personnel to perform their tasks.

- The applicant verified that the design of the HSIs conforms to HFE guidelines (such as the applicant's style guide).

- The applicant validated, using performance-based tests, that the integrated system design (i.e., hardware, software, procedures and personnel elements) supports the safe operation of the plant.

- The applicant has (1) evaluated HEDs to determine if they require corrections, (2) identified design solutions to address HEDs that must to be corrected, and (3) verified the completed implementation of these HED design solutions.

11.3 Applicant Products and Submittals

The product of the applicant's V&V program is a completed design that is verified and validated.

The applicant should provide either an IP or a completed RSR. If the applicant submits an IP, it should describe the complete methodology for conducting V&V, including:

- the inventory developed to characterize the HSIs
- the criteria to be used for Task Support Verification and HFE Design Verification

- the complete set of detailed scenarios for ISV (and how they were identified through the Sampling of Operational Conditions), performance measures, and acceptance criteria

- the methods by which HEDs will be evaluated

The NRC will review this material using the criteria in Section 11.4 below. Then the applicant will submit the RSR when the work described by the IP is completed.

If the applicant submits a completed RSR, the NRC will verify the results using the criteria in Section 11.4 below. At a minimum, the RSR should include the following:

- a description of the methodology, if an NRC approved IP was not used

- a description of the results from Task Support Verification and HFE Design Verification

- details of the results of the ISV, including a statement of how the validation demonstrates the ability to safely operate the plant

- a list of HEDs generated from the V&V, the analyses associated with these HEDs, and their resolutions

Summaries may be used for any of the above items provided that references are given for more detailed documents. If the methodology was described in an IP that the NRC staff previously reviewed, the contents of the RSR should be consistent with the approved methodology and the applicant should discuss the rationale for any deviations from it.

In addition to reviewing the applicant's documentation, the NRC staff also may verify a sample of V&V activities to confirm the results, and observe the integrated-system validation trials.

11.4   Review Criteria

11.4.1 Sampling of Operational Conditions

As stated in Section 11.2, the objective of the Sampling of Operational Conditions review is to verify that the applicant identified a sample of operational conditions that (1) includes conditions representative of the range of events that could be encountered during the plant's operation, (2) reflects the characteristics expected to contribute to variations in the system's performance, and (3) considers the safety significance of HSIs. These sample characteristics are best identified by using a multidimensional sampling strategy to reasonably assure that V&V evaluations include variation along important dimensions.

The sampling methodology will identify a range of operational conditions to guide Task Support Verification, HFE Design Verification, and ISV. The NRC's review of this activity considers the dimensions to be used to identify and select conditions, and their integration into scenarios.

11.4.1.1   Sampling Dimensions

The following sampling dimensions are addressed below: Plant conditions, personnel tasks, and situational factors known to challenge personnel performance.

(1)   The applicant should include the following plant conditions:

- normal operational events including plant startup, shutdown or refueling, and significant changes in operating power

- I&C and HSI failures and degraded conditions that encompass:

  - The I&C system, including the sensor, monitoring, automation and control, and communications subsystems; [e.g., safety-related system logic and control unit, fault tolerant controller, local "field unit" for multiplexer (MUX) system, MUX controller, and a break in MUX line]

  - common cause failure of the I&C system during a design basis accident (as defined by BTP 7-19)

  - HSIs including, loss of processing or display capabilities for alarms, displays, controls, and computer-based procedures

- transients and accidents, such as:

  - transients (e.g., turbine trip, loss of off-site power, station blackout, loss of all feedwater, loss of service water, loss of power to selected buses or MCR power supplies, and safety and relief valve transients)

  - accidents (e.g., main-steam-line break, positive reactivity addition, control rod insertion at power, anticipated transient without scram, and various-sized loss-of-coolant accidents)

  - reactor shutdown and cooldown using the remote shutdown system

  - reasonable, risk-significant, beyond-design-basis events that should be determined from the plant-specific PRA

(2) The applicant should include the following types of personnel tasks:

- *Important HAs, Systems, and Accident Sequences* – The sample should include all important HAs, as determined in Section 7. Additional factors that contribute highly to risk, as defined by the PRA, also should be sampled:

  - dominant accident sequences

  - dominant systems (selected through PRA importance measures, such as Risk Achievement Worth or Risk Reduction Worth)

- *Manual Initiation of Protective Actions* – The sample should include manual system-level actuation of critical safety functions.

- *Automatic System Monitoring* – The sample should include situations in which humans must monitor a risk-important automatic system.

- *OER-Identified Problematic Tasks* – The sample should include all personnel tasks identified as problematic during the applicant's review of operating experience.

- *Range of Procedure Guided Tasks* –The sample should include tasks that are well defined by procedures. Personnel should be able to understand and execute the specified steps as part of their rule-based decision-making. Regulatory Guide 1.33, Appendix A, contains several categories of "typical safety-related activities that should be covered by written procedures." The sample should include appropriate procedures in each category:

  - administrative procedures

76

- general plant operating procedures

- procedures for startup, operation, and shutdown of safety-related systems

- procedures for abnormal, off-normal, and alarm conditions

- procedures for combating emergencies and other significant events (e.g., reactor accidents, and declaration of emergency-action levels)

- procedures for controlling radioactivity

- procedures for controlling measuring and test equipment and for surveillance tests, procedures, and calibration

- procedures for performing maintenance

- chemistry and radiochemical control procedures

- *Range of Knowledge-Based Tasks* – The sample should include tasks that are not well defined by detailed procedures.

  Additional Information: A situation may demand knowledge-based decision-making if the procedural rules do not fully address the problem, or when the selection of an appropriate rule is unclear. An example in a pressurized water reactor plant may be the difficulty in diagnosing a steam generator tube rupture (SGTR) with a failure of radiation monitors on the plant's secondary side. This happens because (1) there is no main indication of the rupture (the presence of radiation in secondary side), and (2) the other effects of the rupture (i.e., slight changes in pressures and levels on the primary and secondary sides) may be attributed to other causes. While the operators may use procedures to treat the symptoms of the event, the determination that the cause is a SGTR may call for a situational assessment based on an understanding of the plant's design and the possible combinations of failures that entail the observed symptoms. Errors in rule-based decision-making result from selecting the wrong rule, or incorrectly applying a rule. Errors in knowledge-based decision-making result from mistakes in higher-level cognitive functions, such as judgment, planning, and analysis. The latter are more likely to occur in complex failure events wherein the symptoms do not resemble the typical case, and thus, are not amenable to pre-established rules.

- *Range of Human Cognitive Activities* – The sample should include the range of cognitive activities that personnel perform, including:

  - detecting and monitoring (e.g., of critical safety-function threats)

  - situation assessment (e.g., interpreting alarms and displays to diagnose faults in plant processes and in automated control and safety systems)

  - planning responses (e.g., evaluating alternatives to recover from plant failures)

  - response implementation (e.g., in-the-loop control of plant systems, assuming manual control from automatic control systems, and carrying out complicated control actions)

  - obtaining feedback (e.g., feedback of the success of actions taken)

- *Range of Human Interactions* – The sample should include the range of interactions among plant personnel, including tasks performed independently by individual crew members, and those undertaken by a team of crew members. These interactions among plant personnel should include interactions between:

  - main control room operators (e.g., operations, shift turnover walkdowns)

- main control room operators with auxiliary operators and other plant personnel performing tasks locally (e.g., maintenance or I&C technicians, chemistry technicians)

- main control room operators and the TSC and the EOF

- main control room operators with plant management, the NRC, and other outside organizations

(3)   The applicant should include the following situational factors or error-forcing contexts known to challenge human performance. It also should include situations specifically designed to create human errors to assess the system's error tolerance, and the ability of personnel to recover from any errors, should these occur, for example:

- *High-Workload Situations* – The sample should include situations where variations in human performance due to high workload and multitasking situations can be assessed.

- *Varying-Workload Situations* – The sample should include situations wherein variations in human performance due to workload transitions can be determined. These include conditions where there is (1) a sudden increase in the number of signals that must be detected and processed after a period in which signals were infrequent, and (2) a rapid reduction in the need for detecting signals and processing demands following a time of high sustained task-demand.

- *Fatigue Situations* – To the extent possible, the sample should include situations that may be associated with fatigue, such as work on backshifts and tasks performed frequently with repetitive actions, such as repeated inputs to a touch screen during plant operations or pulling rods.

- *Environmental Factors* – To the extent possible, the sample should include environmental conditions that may cause human performance to vary, e.g., poor lighting, extreme temperatures, high noise, and simulated radiological contamination.

11.4.1.2   Identification of Scenarios

(1)   The applicant should combine the results of the sampling to identify a set of V&V scenarios to guide subsequent analyses.

Additional Information: A given scenario may combine many of the characteristics identified by sampling of operational conditions.

(2)   The applicant should not bias the scenarios by overly representing the following:

- scenarios for which only positive outcomes are expected

- scenarios that, for ISV, are relatively easy to conduct (i.e., scenarios should not be avoided simply because they are demanding to set up and run on a simulator)

- scenarios that, for ISV, are familiar and well structured (e.g., which address familiar systems and failure modes that are highly compatible with plant procedures, such as "textbook" design-basis accidents)

11.4.1.3 Scenario Definition

(1) The applicant should identify operational conditions and scenarios to be used for HSI Task Support Verification, Design Verification, and ISV. The applicant should develop detailed scenarios suitable for use on a full-scope simulator. The level of detail should be comparable to what one would include in a test plan. For each one, the following information should be defined to reasonably assure that important dimensions of performance are addressed, and to allow the scenarios to be accurately and consistently presented for repeated trials:

- a description of the scenario and any pertinent prior history necessary for personnel to understand the state of the plant at the start-up of the scenario

- specific initial conditions (a precise definition of the plant's functions, processes, systems, component conditions, and performance parameters, e.g., similar to that at shift turnover)

- events (e.g., failures) that will occur during the scenario and their initiating conditions, e.g., based on time, or a value of a specific parameter

- precise definition of workplace factors, (e.g., environmental conditions, such as low levels of illumination)

- needs for task support (e.g., procedures and technical specifications)

- staffing level

- details of communication content between control room personnel and remote personnel (e.g., load dispatcher via telephone)

- scripted responses for test personnel who will act as plant personnel in the test scenarios

  Additional Information: Test personnel act as surrogates for personnel outside the control room. To the greatest extent possible, prepare responses to questions that may be asked by operators communicating with the personnel outside the control room. There are limits to the ability to preplan communications because personnel may ask unanticipated questions or make unforeseen requests. However, efforts should be made to detail what information personnel outside the control room can provide, and script the responses to likely questions.

- the precise specification of what, when, and how data are to be collected and stored (including videotaping, questionnaires, and rating-scale administrations)

- precise specifications on simulator set up

- specific criteria for terminating the scenario

(2) The applicant's scenarios should realistically replicate operator tasks in the tests; then, the findings from the test can be generalized to the plant's actual operations.

(3) When the applicant's scenarios include work associated with operations remote from the main control room, the effects on personnel performance due to potentially harsh environments (e.g., high radiation) should be realistically simulated (e.g., additional time to don protective clothing, and access radiologically controlled areas).

11.4.1.4 Additional Considerations for Reviewing the HFE Aspects of Plant Modifications

In addition to any of the criteria above that relate to the modification being reviewed, the applicant should address the following considerations.

(1)   The applicant's operational conditions should reflect tasks that involve a modification, rather than the entire range of topics discussed in Section 11.4.1.

(2)   For ISV, the applicant's operational conditions should encompass the transfer of learning effects on personnel performance when modifying an old HSI or procedure.

        Additional Information: Negative transfer of learning may occur when the new and old components are different and impose different demands on personnel.

(3)   For ISV, when both old and new versions of the same HSIs are permanently present in the HSI but with different means of presentation and methods of operation, then the applicant's evaluations should reasonably assure that personnel can alternate their use of these HSIs without degrading performance.

(4)   Where old HSIs are to be deactivated but left in place in the HSI, the applicant should identify conditions for an ISV that would test the potential for their interfering with tasks.

        Additional Information: For example, the presence of deactivated HSIs may cause visual clutter that interferes with the ability of personnel to locate and use other HSIs.

11.4.2 Design Verification Review Criteria

11.4.2.1 HSI Inventory and Characterization

As stated in Section 11.2, the objective of the review is to verify that the applicant's HSI inventory and characterization accurately describes all HSI displays, controls, and related equipment lying within the scope defined by the sampling of operational conditions.

Applicants may document their HSI inventory in different ways. They should describe the means by which this is done, and provide it to the NRC's staff for review using the criteria in this section.

(1)   *Scope* – The applicant should develop an inventory of all HSIs that personnel require to complete the tasks covered in the validation scenarios that were identified by the applicant's Sampling of Operational Conditions. The inventory should include aspects of the HSI used for managing the interface, such as navigation and retrieving displays, as well as those that control the plant.

(2)   *HSI Characterization* – The applicant's inventory should describe the characteristics of each HSI within the scope of the verification. The following is a minimal set of information for this characterization:

- a unique identification code number or name
- associated plant system and subsystem
- associated personnel functions and tasks

- type of HSI, e.g.,

    - computer-based control (e.g., touch screen or cursor-operated button and keyboard input)

    - hardwired control (e.g., J-handle controller, button, and automatic controller)

    - computer-based display (e.g., digital value and analog representation)

    - hardwired display (e.g., dial, gauge, and strip-chart recorder)

- display characteristics and functionality [e.g., plant variables/parameters, units of measure, accuracy of variable/parameter, precision of display, dynamic response, and display format (e.g., bar chart or trend plot)]

- control characteristics and functionality [e.g., continuous versus discrete settings, number and type of control modes, accuracy, precision, dynamic response, and control format (method of input)]

- user-system interaction and dialog types (e.g., navigation aids and menus)

- location in data-management system (e.g., identification code for information display screen)

- physical location in the HSI (e.g., control panel section), if applicable

The applicant should include photographs, copies of display screens, or similar samples of HSIs in the HSI inventory and characterization.

(3)   *Inventory Verification* – The applicant should verify the inventory description of HSIs to ensure that it accurately reflects their current state.

11.4.2.2 HSI Task Support Verification

HSI Task Support Verification addresses the availability of items needed to support task requirements. As stated in Section 11.2, the objective of the HSI Task Support Verification review is to ensure that the applicant verified that the HSI provides the needed alarms, information, controls, and task support for personnel to perform their tasks, defined by the task analysis.

(1)   *Verification Criteria* – The applicant should base the HSI task support criteria on the alarms, controls, displays, and task support needed by personnel to complete their tasks as identified by the applicant's task analysis.

(2)   *General Methodology* – The applicant should compare the HSIs and their characteristics (as defined in the HSI inventory and characterization) to the needs of personnel identified in the task analysis for the defined sampling of operational conditions, noted in Section 11.4.1.

(3)   *HED Identification* – The applicant should identify and document an HED when:

- An HSI needed for task performance (e.g., a necessary control or display) is unavailable.

- HSI characteristics do not match the requirements of the personnel task (e.g., a display may show the needed plant parameter but not within the range or precision needed for the task).

- HSIs are available that are not needed for any task.

Additional Information: Unnecessary HSIs introduce clutter, and can distract personnel from selecting the appropriate ones. It is important to verify that the HSI is unnecessary. Appropriate ones may not appear to be needed with personnel tasks for the following reasons:

- The HSI is essential for a task that the task analysis did not address (i.e., it was not within the scope of the design review).

- The task analysis was incomplete, overlooking the need for the HSI.

- The HSI only partially meets the established requirements for the personnel task.

(4) *HED Documentation* – The applicant should document HEDs to identify the HSI, the tasks affected, and the basis for the deficiency (what aspect of the HSI was identified as not meeting task requirements).

Additional Information: The analysis and correction of HEDs is detailed in Section 11.4.4, Human Engineering Discrepancy Resolution Review Criteria.

(5) *Additional Methodology Considerations for Plant Modifications* – In addition to any of the criteria above that relate to the modification being reviewed, the applicant should address the following considerations:

- HSI Task Support Verification should address all aspects of HSIs described above related to the modification. For modifications to plant systems that do not include modifications of the HSIs, verification of task support should highlight any new demands for monitoring and control, and assess whether the existing HSI design adequately addresses them.

- HSI Task Support Verification should cover configurations in the modification in which old HSIs are deactivated permanently, but not removed (e.g., abandoned in place). Criterion 4 in this subsection states that the HSIs should not contain any information, displays, or controls that do not support personnel tasks. This verification should identify deactivated HSIs that might negatively affect personnel performance, such as obstructing the view of important information or adding visual clutter that could interfere with monitoring. The applicant should identify deactivated HSIs requiring further evaluation through HFE design verification or ISV.

- HSI Task Support Verification should address the temporary configurations of the HSIs and plant systems that may be created when establishing the modification, and so used by operations and maintenance personnel when the plant is not shutdown. These configurations may include:

  - the use of HSIs that differ from the intended final design

  - combinations of HSIs and system configurations that differ from both the original design and the intended final one

  For each temporary HSI configuration, the task requirements of personnel should be identified and compared to the information and control capabilities available.

Additional Information: For example, if a temporary configuration of plant systems introduces special monitoring requirements, the HSIs should provide the necessary information.

11.4.2.3 HFE Design Verification

HFE Design Verification addresses the suitability of the HSI with regard to human capabilities and limitations. As stated in Section 11.2, the objective of the HFE Design Verification review is to evaluate the applicant's verification that the design of the HSIs conforms to HFE guidelines.

(1)   *Verification Criteria* – The applicant should base the criteria used for HFE Design Verification on HFE guidelines.

Additional Information: The choice of guidelines used in this verification depends upon whether the applicant developed a design-specific style guide. The acceptability of the style guide used by the applicant should be reviewed by the NRC staff using the review guidance in Section 8.4.3, HFE Design Guidance for HSIs. Using an NRC-reviewed style guide affords the criteria for verifying the HFE design. When no style guide is available, the guidelines in NUREG-0700 can be used by the applicant for this purpose. However, because not all of the guidelines therein will be applicable to each review, the applicant should select those based on the characteristics of the HSIs being evaluated. Applicants should identify a subset of guidelines appropriate to a specific design based on the HSI characterization.

(2)   *General Methodology* – The applicant's HFE Design Verification methodology should include the following:

- Procedures for comparing the characteristics of the HSIs with HFE guidelines for (1) the defined sampling of operational conditions, as noted in Section 11.4.1, and (2) the general environment in which HSIs are sited, including workstations, control rooms, and environmental characteristics (e.g., lighting and noise).

  Additional Information: A single guideline may apply to many HSIs. By verifying all HSIs within the scenarios defined in Section 11.4.1, the consistency of applying a guideline across multiple HSIs can be assessed.

- Procedures for determining for each guideline whether the HSI is "acceptable" or "discrepant." If discrepant, it should be designated as an HED, tracked, and evaluated (see Sections 2.4.4 and 11.4.4).

  Additional Information: A judgment that an HSI is "acceptable" should be made only if compliance is total, i.e., only if every instance of the item is fully consistent with the criteria established by the HFE guidelines. If there is any noncompliance, full or partial, then an evaluation of "discrepant" should be given, and a notation made as to where it occurs.

- Procedures for evaluating whether an HED is a potential indicator of additional issues.

  Additional Information: For example, identifying an inappropriate format for presenting data on an individual display should be considered a potential sign that other display formats might be used incorrectly, or that the observed format is employed inappropriately elsewhere. Then, the sampling strategy should be modified to encompass other display formats. In some cases, discovering these discrepancies will warrant further review in the identified areas of concern.

(3)   *HED Identification* – The applicant should identify an HED when a characteristic of the HSI is "discrepant" from a guideline.

(4)   *HED Documentation* – The applicant should document HEDs in terms of the HSI involved, and how its characteristics depart from a particular guideline.

> *Additional Information*: The analysis and correction of HEDs is addressed in Section 11.4.4, Human Engineering Discrepancy Resolution Review Criteria.

(5)   *Additional Considerations for Reviewing the HFE Aspects of Plant Modifications* – In addition to any of the criteria above that relate to the modification being reviewed, the applicant should address the following considerations:

- The scope of HFE design verification may be restricted to the modified HSIs and their interactions with the rest of the HSIs.

- When both old and new versions of similar HSIs are available, this verification should offer reasonable assurance that their means of presentation and methods of operation are compatible, such that personnel performance will not be impaired when alternating the use of each one.

- HEDs should be identified for the following:
    - failure to meet "personnel-identified" functionality in addition to that specified by system designers. When a digital system replaces an existing system, it is important to ensure that all operational uses of the former system were addressed, even those that were not intended in the original design. The replacement system's design should consider the ways in which personnel actually used the former system
    - poor integration with the rest of the HSI
    - poor integration with procedures and training

- Temporary configurations of the HSIs and plant systems that operations and maintenance personnel may use when the plant is not shutdown, should be reviewed to verify that their design is consistent with the principles of good HFE design, including consistency with the rest of the HSIs.

11.4.3 Integrated System Validation

As stated in Section 11.2, the objective of the ISV review is to verify that the applicant validated, using performance-based tests, that the integrated system design (i.e., hardware, software, procedures and personnel elements) supports the safe operation of the plant.

The scenarios for ISV should be performed using a simulator, or other suitable representation of the system, to determine the complete design's adequacy to support safe operations. Validation should be performed after the resolution of all significant HEDs identified in verification reviews.

Applicants submitting an ISV IP for staff review should follow the general guidance in Section 1.2.2. The IP should describe the methodology of the tests that will be performed. It should identify the specific scenarios to be used, and detail them at a level that will support the staff's review, using the criteria stated in this section. The level of scenario detail should be

84

comparable to that in a test plan. For each scenario, the applicant should specify the specific performance measures used for pass/fail along with the criteria for diagnostic evaluations to be used in assessing the results. The NRC will not accept submittals that merely provide a plan for developing the detailed ISV methodology.

The applicability and scope of the ISV may vary in reviewing the HFE aspects of plant modifications. An ISV should be reviewed for all modifications that may (1) change personnel tasks; (2) change tasks demands, such as changing the task's dynamics, complexity, or workload; or, (3) interact with or affect HSIs and procedures in ways that may degrade performance. ISV may not be needed when a modification involves only minor changes to personnel tasks such that the modification reasonably may be expected to have little or no overall effect on workload and the likelihood of error. Those aspects of validation that should be addressed in the NRC staff's evaluation are discussed below.

11.4.3.1 Validation Team

(1)   The applicant should describe how the team performing the validation has independence from the personnel responsible for the actual design.

> *Additional Information*: The members of the validation team should have no responsibility for the design; i.e., they should never have been part of the design team. While they may work for the same organization, their responsibilities must not include contributions to the design, other than validating it.

11.4.3.2 Test Objectives

(1)   The applicant should develop detailed test objectives to provide evidence that the integrated system adequately supports plant personnel in safely operating the plant, to include the following considerations:

- Validate the acceptability of the shift staffing level(s), the assignment of tasks to crew members, and crew coordination within the control room, between the control room and local control stations and support centers, and with individuals performing tasks locally. This should encompass validating minimum shift staffing levels, nominal levels, maximum levels, and shift turnover (see Section 6 for definitions).

- Validate that the design has adequate capability for alerting, informing controlling, and feedback such that personnel tasks are successfully completed during normal plant evolutions, transients, design-basis accidents, and also under selected, risk-significant events beyond-design basis, as defined by sampling operational conditions.

- Validate that specific personnel tasks can be accomplished within the time and performance criteria, with effective situational awareness, and acceptable workload levels that balance vigilance and personnel burden.

- Validate that the HSIs minimize personnel error and assure error detection and recovery capability when errors occur.

- Validate the assumptions about performance on important HAs.

> *Additional Information*: For example, the HRA within the plant PRA contains several assumptions regarding the performance of risk-important HAs. These assumptions should be validated for dominant sequences, such as decision-making and diagnosis strategies, and

also for the human actions. This process should be completed before the final quantification stage of the PRA.

- Validate that the personnel can effectively transition between the HSIs and procedures in accomplishing their tasks, and that interface management tasks, such as display configuration and navigation, are not a distraction or an undue burden.

(2) *Additional Considerations for Reviewing the HFE Aspects of Plant Modifications* – In addition to any of the criteria above that relate to the modification being reviewed, the test's objectives and scenarios should be developed to encompass aspects of performance affected by the modified design (even when the HSIs are not modified), including personnel tasks.

11.4.3.3 Validation Testbeds

A testbed is the HSI representation used to perform validation evaluations. One approach an applicant can use to acceptably meet criteria 1 through 7 in this section is to use a testbed that is compliant with "Nuclear Power Plant Simulators for Use in Operator Training" (ANS, 2009).

(1) *Interface Completeness* – The applicant's testbed should represent completely the integrated system. It should include HSIs and procedures not specifically required in the test scenarios.

 Additional Information: Adjacent controls and displays may affect the ways in which personnel use those addressed by a particular validation scenario.

(2) *Interface Physical Fidelity* – The testbed's HSIs and procedures should be represented with high physical fidelity to the reference design, including the presentation of alarms, displays, controls, job aids, procedures, communications equipment, interface management tools, layout, and spatial relationships.

(3) *Interface Functional Fidelity* – The testbed's HSI and procedure functionality should be represented with high fidelity to the reference design. All HSI functions should be available.

 Additional Information: High fidelity covers the HSI modes of operation (i.e., the changes in functionality that can be invoked by personnel selecting them), or changes in plant states.

(4) *Environmental Fidelity* – The testbed's environmental fidelity should be represented with high physical fidelity to the reference design, including the expected levels of lighting, noise, temperature, and humidity. Thus, for example, the noise contributed by equipment, such as air-handling units, computers, and communications equipment should be represented in validation tests.

(5) *Data Completeness Fidelity* – Information and data provided to personnel should completely represent the plant's systems they monitor and control.

(6) *Data Content Fidelity* – The testbed's data content fidelity should be represented with high physical fidelity to the reference design. The presentation of information and controls should rest on an underlying model accurately mirroring the reference plant. The model should provide input to the HSI such that the information accurately matches that which is presented during operations.

(7) *Data Dynamics Fidelity* – The testbed's data dynamics fidelity should be represented with high fidelity to the reference design. The process model should be able to provide input to the HSI so that information flow and control responses occur accurately and within the correct response time; e.g., information should be sent to personnel with the same delays as occur in the plant.

(8) For important HAs at complex HSIs remote from the main control room (e. g., a remote shutdown facility), where timely, precise actions are essential, the use of a simulator or mockup should be considered to verify that the requirements for human performance can be met. (For less important HAs, or for non-complex HSIs, human performance may be assessed on analysis, such as task analysis, rather than on simulations.)

(9) The applicant should verify the conformance of the testbed to the testbed-required characteristics before validation tests are conducted.

11.4.3.4 Plant Personnel

(1) Participants in the applicant's validation tests should be representative of plant personnel who will interact with the HSI (e.g., licensed operators, rather than training personnel or engineers).

(2) To properly account for human variability, the applicant should use a sample of participants that reflects the characteristics of the population from which it is drawn. Those characteristics expected to contribute to variations in system performance should be specifically identified; the sampling process should reasonably assure that the validation encompasses variation along that dimension. Determining representativeness should include considering the participants' license type and qualifications, skill/experience, age, and general demographics.

(3) In selecting personnel for participating in the tests, the applicant should consider the minimum shift staffing levels, nominal levels, and maximum levels, including shift supervisors, reactor operators, shift technical advisors, etc.

(4) The applicant should prevent bias in the sample of participants by avoiding the use of participants who:

- are members of the design organization

- participated in prior evaluations

- were selected for some specific characteristic, such as crews identified as good performers or more experienced

11.4.3.5 Performance Measurement

ISV employs a hierarchal set of performance measures including measures of plant performance, personnel task performance, situation awareness, cognitive workload, and anthropometric/physiological factors. Errors of omission and commission also are identified. A hierarchal set of measures provides sufficient information to validate the integrated system design and affords a basis to evaluate deficiencies in performance and thereby identify needed improvements. Pass/fail measures are those used to determine whether the design is or is not validated. Diagnostic measures are used to better understand personnel performance and to facilitate the analyses of errors and HEDs.

11.4.3.5.1   Types of Performance Measures

(1)   The applicant should identify the specific plant performance measures applicable to each ISV scenario.

Additional Information: They may address the performance of functions, systems, or components.

(2)   The applicant should identify the primary task measures applicable to each ISV scenario.

- For each scenario, the applicant should identify the primary tasks operators must perform to accomplish scenario goals, so that such measures can be developed.

    Additional Information: The primary tasks are those involved in carrying out the functional role of the operator in supervising the plant; i.e., monitoring, detection, situation assessment, response planning, and response implementation. Primary tasks should be assessed at a level of detail appropriate to the task's demands. For example, for some simple scenarios, measuring the time to complete a task may suffice. For complicated tasks, especially those described as knowledge-based, it may be appropriate to undertake a fine-grained analysis, such as identifying the task's components, viz., seeking specific data, making decisions, taking actions, and obtaining feedback.

- The measures chosen to evaluate personnel task performance should reflect those aspects of the task that are important to system performance, such as:

    - time
    - accuracy
    - frequency
    - amount achieved or accomplished
    - consumption or quantity used
    - subjective reports of participants
    - behavior categorization by observers

- The analysis of primary tasks will support the identification of errors of omission (primary tasks not performed). Also, any actions and tasks that operators *actually* perform that deviate from the primary tasks should be identified and noted. These actions should be used to identify errors of commission.

(3)   The applicant should identify the secondary task measures applicable to each scenario.

Additional Information: Secondary tasks are those personnel must perform when interfacing with the HSI, such as navigating through computer screens to find a needed display and to configure HSIs. The measurement of secondary task performance should reflect the demands of the detailed HSI implementation, e.g., time to configure a workstation, navigate between displays, and manipulate them (e.g., changing display type and scale settings).

(4)   The applicant should identify the measures of situation awareness applicable to each scenario.

    Additional Information: Situation awareness is the degree to which personnel's perception of plant parameters and understanding of the plant's condition corresponds to its actual condition at any given time and influences predictions about future states.

(5)   The applicant should identify the workload measures obtained for each scenario.

    Additional Information: Workload is comprised of the physical, cognitive, and other demands that tasks place on plant personnel. The impact of one or many of these aspects of workload should be considered in the performance measures.

(6)   The applicant should identify the anthropometric and physiological measures obtained for each scenario.

    Additional Information: Anthropometric and physiological factors include such concerns as visibility of displays, accessibility of control devices, and ease of manipulating the control device. Many of these design aspects are assessed as part of verifying the HFEs design. Therefore, attention should focus on those areas of the design that only can be addressed by testing the integrated system, e.g., the ability of personnel effectively to use the various controls, displays, workstations, or consoles while performing their tasks.

11.4.3.5.2   Performance Measure Information and Validation Criteria

(1)   The applicant should describe the methods by which these measures are obtained, e.g., by simulator data recording, participant questionnaires, or observation by subject-matter experts.

(2)   The applicant should specify when each measure is obtained (recorded), such as continuously, at specific points during the scenario, or after the scenario ends.

(3)   The applicant should describe the characteristics (see Table 11-1) of the performance measures.

Table 11-1 Characteristics of Performance Measures

Characteristic	Meaning
Construct Validity	A measure should represent accurately the aspect of performance it is intended to measure.
Reliability	A measure should be repeatable; i.e., same behavior measured in exactly the same way under identical circumstances should yield the same results.
Sensitivity	A measure's range (scale) and its frequency (how often data are collected) should be appropriate to that aspect of performance being assessed.
Unobtrusiveness	A measure should minimally alter the psychological or physical processes that are being investigated.
Objectivity	A measure should be based on easily observed phenomena.

(4)   The applicant should identify the specific criterion for each measure used to judge the acceptability of performance and describe its basis.

Additional Information: Table 11-2 describes the different bases for performance criteria.

Table 11-2 Basis for Performance Criteria

Criteria Basis	Meaning
Requirement	The observed performance of the integrated system is compared with a quantified performance requirement; i.e., the requirements for the performance of systems, subsystems, and personnel are defined through engineering analyses.
Benchmark	The observed performance of the integrated system is compared with a criterion established using a benchmark system, e.g., a current system is predefined as acceptable.
Norm	The observed performance of the integrated system is compared with a criterion using many predecessor systems (rather than a single benchmark system).
Expert Judgment	The observed performance of the integrated system is compared with a criterion established by subject-matter experts.

(5)   The applicant should identify whether each measure is a pass/fail one or a diagnostic one.

11.4.3.6   Test Design

The criteria in this section are divided into the following subsections:

> 11.4.3.6.1, Scenario Sequencing
> 11.4.3.6.2, Test Procedures
> 11.4.3.6.3, Test Personnel Training
> 11.4.3.6.4, Participant Training
> 11.4.3.6.5, Pilot Testing

11.4.3.6.1   Scenario Sequencing

(1)   The applicant should balance scenarios across crews to provide each crew with a similar, representative range of scenarios.

Additional Information: Random assignment of scenarios to crews for ISV is undesirable. The value of using random assignment to control bias is effective only when the number of crews is quite large.

(2)   The applicant should balance the order of presentation of scenarios to crews to provide reasonable assurance that the scenarios are not always presented in the same sequence (e.g., the easy scenario is not always used first).

11.4.3.6.2   Test Procedures

(1)   The applicant should use detailed, unambiguous procedures to govern the conduct of the tests. These procedures should include the following:

• the identification of which crews receive which scenarios, and the order in which they should be presented

• detailed and standardized instructions for briefing the participants

Additional Information: The type of instructions given to participants can affect their performance on a task. This source of bias is minimized by developing standard instructions.

- specific directions for the testing personnel on conducting the test scenarios, as elaborated in Scenario Definition (Section 11.4.1.3)

- guidance on when and how to interact with participants when difficulties occur in simulation or testing

  Additional Information: Even when a high-fidelity simulator is used, the participants may encounter artifacts of the test environment that detract from their performance of the tasks that are the focus of the evaluation. Guidance should be available to the test conductors to help resolve such conditions.

- instructions on when and how to collect and store data. These instructions should stipulate which data are to be recorded by:

  - simulator computers

  - special-purpose instruments and devices for collecting data (such as situation awareness- and workload-questionnaires, or physiological measures)

  - video recorders (locations and views)

  - test personnel and subject-matter experts (such as via observational checklists)

- procedures for documentation:

  - identifying and maintaining files of test records including details of the crew and scenarios

  - data collected

  - logs created by those who conducted the tests

  The procedures should detail the types of information that should be logged (e.g., when the tests were performed, deviations from the test procedures and why they occurred, and any unusual events that may be important to understanding how a test was run or for interpreting the findings from it). The procedure also should state when the types of information should be recorded.

(2) The applicant's test procedures should minimize the opportunity for bias in the test personnel's' expectations and in the participant's responses.

Additional Information: The expectancies of test personnel may introduce a bias if the expectations of the testers systematically influence the collection of data. Expectancies can influence performance in many ways (e.g., test personnel may, by giving subtle cues or communications, provide direction to participants, or they may tend to evaluate the performance of participants in ways that reflect more favorably upon the design than would an objective observer). Participant response bias means that the design of the test itself affects the data obtained from participants. It is not necessarily implied that a response bias represents any deliberate attempt by the participants to be untruthful. The test environment can influence participants in ways that have little to do with the tests objectives. Response bias can occur in four ways. First, participants may wish to influence outcomes and so be biased toward producing data consistent with their desired result. Second, participants may want to provide data that they think the test personnel want to obtain. Third, participants may try to figure out how performance should vary under different conditions, and then influence data to be consistent with such differences. Fourth, participants may want to excel because they know that they are being observed. See NUREG/CR 6393 (O'Hara et al., 1997) for additional information.

11.4.3.6.3 Training Test Personnel

(1)   The applicant should train test personnel (those who conduct or administer the validation tests) on the following:

- the use and importance of test procedures
- bias and errors that test personnel may introduce into the data through failures to follow test procedures accurately or to interact with participants properly
- the importance of accurately documenting problems arising during testing, even if they were due to an oversight or error of those conducting the test

11.4.3.6.4 Training Participants

(1)   The applicant's training of participants should be very similar to the training plant personnel receive. It should reasonably assure that the participants' knowledge of the plant's design, and operations, and the use of the HSIs and procedures represents that of experienced plant personnel. Participants should not be trained specifically to carry out the selected validation scenarios.

(2)   To assure that the participants' performance is representative of plant personnel, the applicant's training of participants should result in near asymptotic performance (i.e., stable, not significantly changing from trial to trial) and should be tested for such before conducting the validation.

11.4.3.6.5 Pilot Testing

(1)   The applicant should conduct a pilot study before the validation tests begin to offer an opportunity for the applicant to assess the adequacy of the test design, performance measures, and data-collection methods.

(2)   The applicant should not use participants in the pilot testing who will then be participants in the validation tests.

11.4.3.7 Data Analysis and HED Identification

(1)   The applicant should use a combination of quantitative and qualitative methods to analyze data. The analysis should reveal the relationship between the observed performance and the established performance criteria.

(2)   The applicant should discuss the method by which data is analyzed across trials, and include the criteria used to determine successful performance for a given scenario.

(3)   The applicant should evaluate the degree of convergence between related measures (i.e., consistency between measures expected to assess the same aspect of performance).

Additional Information: For example, if situation assessment is measured by both a participant questionnaire, and an observer rating scale, the results should be consistent with each other. If they do not converge, the reason for this should be identified.

(4)   When interpreting test results, the applicant should allow a margin of error to reflect the fact that actual performance may be slightly more variable than observed validation-test performance.

(5)   The applicant should verify the correctness of the analyses of the data. This verification should be done by individuals or groups other than those who performed the original analysis, but may be from the same organization.

(6)   The applicant should identify HEDs when the observed performance does not meet the performance criteria.

Additional Information: The analysis and correction of HEDs is addressed in Section 11.4.4, Human Engineering Discrepancy Resolution Review Criteria.

(7)   The applicant should resolve HEDs identified by pass/fail measures before the design is accepted.

11.4.3.8   Validation Conclusions

(1)   The applicant should document the statistical and logical bases for determining that performance of the integrated system is, and will be acceptable.

(2)   The applicant should document the limitations in the validation tests, their possible effects on the conclusions of the validation, and their impact on implementing the design.

Additional Information: Examples of possible limitations include:

* aspects of the tests that were not well controlled

* potential differences between the test situation and actual operations, such as the absence of productivity-safety conflicts

* potential differences between the validated design and the as-built plant or system (if validation is directed to a plant under construction where such information is available, or to a new design using the validation findings from a predecessor)

11.4.4 Human Engineering Discrepancy Resolution Review Criteria

HEDs are identified in the V&V process during:

* Task Support Verification (Section 11.4.2.2, criterion 3)

* HFE Design Verification (Section 11.4.2.3, criterion 3 and criterion 5 for plant modifications)

* ISV (Section 11.4.3.5.2, criterion 4)

As stated in Section 11.2, the objectives of the NRC staff's review is to verify that the applicant has (1) evaluated HEDs to determine if they require correction, (2) identified design solutions to address HEDs that must be corrected, and (3) verified the completed implementation of these HED design solutions. The applicant's resolution of HEDs is reviewed by the NRC staff using the guidance is this section.

HED Resolution can be performed iteratively throughout V&V. Thus, issues identified during one V&V activity can be addressed and resolved before starting another.

(1) *HED Analysis* – The applicant's HED analyses should include the following:

- *Personnel Tasks and Functions* – The impact of HEDs on personnel tasks and the functions supported by those tasks.

  Additional Information: The potential effects of HEDs is determined, in part, by the importance of the personnel function to plant safety (e.g., consequences of failure), and their cumulative effect on personnel performance (e.g., degree of impairment and types of potential errors).

- *Plant Systems* – The impact of HEDs on plant systems, considering the safety significance of that system(s), their effect on accident analyses, and their relationship to risk-significant sequences in the plant's PRA.

  Additional Information: The potential effects of these HEDs on the plant's safety and personnel performance are determined, in part, by the safety significance of the plant system(s) related to the particular component.

- *Cumulative Effects of HEDs* – The analysis of HEDs should identify the cumulative effects that multiple HEDs may have on plant safety and personnel performance.

  Additional Information: Although an individual HED might not be considered sufficiently severe to warrant correction, the combined effect of several of them on a single aspect of the design could significantly degrade plant safety, and therefore, necessitate corrective action. Likewise, when a single plant system with multiple associated HEDs affects several HSIs, then their possible combined effect on the operation of that plant system should be considered.

- *HEDs as Indications of Broader Issues* – As well as addressing specific HEDs, the applicant's analysis should determine whether the HEDs point to potentially broader problems.

  Additional Information: For example, identifying multiple HEDs associated with one particular aspect of the HSI design, such as the remote shutdown panel, also might suggest other problems with that aspect of the design, such as inconsistent use of design procedures and style guides. In some cases, findings from evaluating HEDs could warrant further review in the identified areas of concern, e.g., when multiple cases of mislabeling are found, the reviewers may wish to do a more complete examination of labeling.

(2) *Selection of HEDs to Correct* – The applicant should conduct an evaluation to identify which HEDs to correct. The evaluation should identify those HEDs that are acceptable as is (The *Additional Information* below provides examples). The remaining discrepancies should be denoted as HEDs to be addressed by the HED-resolution process.

HEDs the applicant should correct are those with direct safety consequences, namely, those that could adversely impact personnel performance such that the margin of plant safety may be reduced below an acceptable level. Unacceptability is indicated by such conditions as violations of Technical Specification safety limits, operating limits, or limiting conditions for operations, or failing an ISV pass/fail criterion.

HEDs with potential safety impact, not as severe as those described above, also should be corrected unless the applicant justifies leaving the condition as is.

The applicant should correct HEDs that may adversely impact personnel performance in a way that has potential consequences to plant performance or SSC operability, and personnel performance or efficiency. This may include failing to meet personnel information needs or violating HFE guidelines for tasks associated with plant productivity, availability, and protecting investment.

Additional Information: HEDs could be acceptable within the context of the fully integrated design. The technical basis for such a determination could include an analysis of recent research literature, current practices, tradeoff studies, or design engineering evaluations.

(3)   *Development of Design Solutions* – The applicant should identify design solutions to correct HEDs. As part of the design solution, the application should evaluate the interrelationships of individual HEDs.

Additional Information: HEDs should not be considered in isolation and to the extent possible, their potential interactions should be considered when developing and implementing solutions. For example, if the HSI for a single plant system is associated with many HEDs, then the set of design solutions should be coordinated to enhance overall performance and avoid incompatibilities between individual solutions. Similarly, if a single plant system is associated with multiple HSIs associated with HEDs, then the design of individual solutions should be harmonized so that the outcome enhances rather than detracts from that system's operation. Approaches that develop design solutions to some HEDs before all are identified in a particular V&V activity are acceptable provided that the potential interactions between HEDs are specifically considered before implementing the design solutions.

(4)   *Design Solution Evaluation* – The applicant should evaluate design solutions to demonstrate the resolution of that HED and to ensure that new HEDs are not introduced. Generally, the evaluation should use the V&V method that originally detected the HED.

Additional Information: For example, if the HED was identified using HFE Design Verification, then that verification should be employed to evaluate the solution. However, there may be reasons for documenting a satisfactory resolution using other methods. For example, if an aspect of the HSI was significantly changed from the resolution of multiple HEDs, the final HSI design may be validated to ensure that the net effect of all the changes is acceptable.

(5)   *HED Evaluation Documentation* – The applicant should document each HED, including:

- the basis for not correcting an HED

- related personnel tasks and functions

- related plant systems

- cumulative effects of HEDs

- HEDs as indications of broader issues

Additional Information: Some, or all, of this documentation may be included in the issues tracking system (Section 2.4.4). Other information, such as cumulative effects or indications of broader issues, may be documented separately.

11.5 Bibliography

ANSI/AIAA G-035A-2000: *Guide to Human Performance Measurements* (AIAA, 2001).

ANSI/ANS 3.5-2009: *Nuclear Power Plant Simulators for Use in Operator Training* (American Nuclear Society, 2009).

IEC 1771: *Nuclear Power Plants Main Control Rooms-Verification and Validation of Design* (International Electrotechnical Commission, 1995).

IEC 60964: *Nuclear Power Plants-Control Rooms-Design* (International Electrochemical Commission, 2009).

IEEE Std. 845-1999: *IEEE Guide to the Evaluation of Human-System Performance in Nuclear Power Generating Stations* (Institute of Electrical and Electronics Engineers, 1999).

ISO 11064-7 *Ergonomic Design of Control Centres: Part 7: Principles for the Evaluation of Control Centres* (ISO, 2006).

NUREG/CR-6393: *Integrated System Validation: Methodology and Review Criteria* (O'Hara et al., 1997).

NUREG-0700: *Human-System Interface Design Review Guidelines* (NRC, 2002).

12   DESIGN IMPLEMENTATION

12.1   Background

This section addresses the implementation (installation and testing) of the HFE aspects of the plant's design for both new plants and plant modifications. For a new plant, the implementation phase is well defined and carefully monitored by start-up procedures and testing. Section 12.4.1 contains the review criteria for new plants.

Implementing HSI modifications is addressed in the plant's overall design modifications program, and may be more complex because aspects of the control room are left as is while others are upgraded. Section 12.4.2 contains the review criteria for modifications.

Plant modifications affect personnel in various ways. Changes to systems and components can impact their role and the way their tasks are performed. Modifications often lead to changes in HSIs, procedures, and training, as well as in the physical equipment. Furthermore, modifications also may involve the HFE aspects of the plant (e.g., the main control room), even though the plant's systems and components are unchanged.

Modifications are implemented in many different ways. Some approaches and their advantages and disadvantages are given in Table 12-1; each has particular HFE considerations.

Table 12-1 Typical Advantages and Disadvantages of Different Methods of Modernization Program Implementation

Many Small Modifications
Advantages
• Minimal disruption to operations.
Potential Disadvantages
• Risk of unexpectedly affecting plant operation (such as through spurious actuation). This could be a problem both for operating and shutdown plants, but potentially more serious for the former.
• Likelihood increases for inconsistency and lack of standardization of HSIs as many new, different systems are added separately to the control room (or other operations and support centers). Consequently, personnel may be unsure precisely how each HSI functions.
• Overlapping functionality; many HSIs are available for personnel to take the same actions.
• Training on small modifications may be lacking, so personnel do not use the new systems effectively or at all.
Large Modifications During a Single Outage
Advantages
• There is no potential for negative effects on personnel performance of interim configurations because the changes all are made at once.
• More economical than multiple outages because (1) interim periods do not have to be analyzed, (2) procedures do not have to be temporarily modified, and (3) personnel do not have to be trained for temporary plant configurations and HSIs.
Potential Disadvantages
• Significant changes to the plant and HSIs can greatly affect the way personnel operate the plant.
Large Modifications During Multiple Outages
Advantages
• Large changes to operations can be minimized by breaking up modifications into smaller logical units.
• Plant staff can gain experience with non-safety systems (less critical), so when safety (critical) systems are modified, the plant's staff already are familiar with the HSIs.

Potential Disadvantages
• Task performance can be hampered if the interim configuration requires parts of a task to be performed using the old HSI, and other parts with the new HSI.
• Interim stages between old- and new-systems especially are error prone if not fully addressed in analyses, and by training and procedural modifications.

Both Old and New Equipment are Left in Place

Advantages
• Any problems with the new system can be identified and resolved while the old HSIs are in place to serve as backups.
• Operators can become familiar with the new HSIs while the old HSIs still are available.
• Old HSIs are available in an emergency (research demonstrated that personnel often prefer the familiar HSIs under stressful conditions).
Potential Disadvantages
• HSI conflicts between old and new systems (such as different values for the same process parameter).
• Control room clutter and potential distraction from two sets of HSIs.
• Different individuals may prefer to the old or the new HSIs, which may adversely impact teamwork.

New Non-functional HSIs in Place in Parallel with Old Functional HSIs

Advantages
• Operators can become familiar with the new HSIs while the old HSIs still are available.
Disadvantages
• Control room clutter and potential distraction of two sets of HSIs.
• Personnel may use the new HSIs inadvertently, or because they do not realize that they are non-functional.

For both new and modified designs, it is important that the applicant determine that the implemented design (i.e., the "as-built" design) accurately reflects the verified and validated design.

12.2   Objective

The NRC staff uses the review criteria in this section to verify that:

• the applicant's as-built design conforms to the verified and validated design resulting from the HFE design process

• the applicant's implementation of plant changes considers the effect on personnel performance, and provides the necessary support to provide reasonable assurance of safe operations

12.3   Applicant Products and Submittals

The product of the applicant's Design Implementation is a final verified and validated as-built HFE design.

The applicant should provide either an IP or a completed RSR. If the applicant submits an IP, it should describe the methodology for conducting design implementation. The NRC will review it using the criteria in Section 12.4 below. Then the applicant will submit the RSR when the work described by the IP is completed.

If the applicant submits a completed RSR, the NRC will verify the results using the criteria in Section 12.4 below. At a minimum, the RSR should include the following:

• describe how the design meets the general criteria in Section 12.4.1

- explain how all aspects of the design that were not addressed during the V&V activities were covered in implementing the design

- document the applicant's verification and concluding statement that the as-built plant conforms to the approved, validated design

- corroborate that all HEDs have been satisfactorily resolved

- delineate how the HFE program addressed each important HA

Summaries may be used for any of the above items provided that references are given for more detailed documents. If the methodology was described in an IP that the NRC staff previously reviewed, the contents of the RSR should be consistent with the approved methodology and the applicant should discuss the rationale for any deviations from it.

12.4 Review Criteria

12.4.1 Final HFE Design Verification for New Plants and Control Room Modifications

(1) The applicant should evaluate aspects of the design that were not addressed in V&V by an appropriate V&V method.

Additional Information: Aspects of the design addressed by this criterion may include design characteristics, such as new or modified displays for plant-specific design features.

(2) The applicant should compare the final HSIs, procedures, and training with the detailed description of the design to verify that they conform to the planned design resulting from the HFE design process and V&V activities. This verification should compare the actual HSI, procedures, and training materials to design descriptions and documents. Any identified discrepancies should be corrected, or justified.

Additional Information: Final design means the design existing in the actual plant.

(3) The applicant should verify that all HFE-related issues in the issue-tracking system (Section 2.4.4) are adequately addressed.

(4) The applicant should provide a description of how the HFE program addressed each important HA.

12.4.2 Additional Considerations for Reviewing the HFE Aspects of Control Room Modifications

In addition to any of the criteria above that are relevant to the modification being reviewed, the following should be addressed.

12.4.2.1 General Criteria for Plant Modifications

(1) The applicant should provide reasonable assurance that the reactor fuel is safely monitored during the shutdown period while physical modifications to the control room are being made.

(2) The applicant should verify that modifications in the plant's procedures and training reflect changes in plant systems, personnel roles and responsibilities, and in HSIs resulting from the new systems.

(3)   Installation should be planned to minimize disruptions to work of plant personnel.

(4)   The applicant should verify that operations and maintenance personnel are fully trained and qualified to operate and maintain all modifications made to the plant before starting-up with the new systems and HSIs in place.

(5)   The applicant should have a plan to monitor startup and initial operations after the modification to reasonably assure that:

- operational and maintenance problems arising from personnel's interactions with the new systems, HSIs, and procedures are identified and addressed

- personnel are sufficiently familiar with the new systems, HSIs, and procedures to support safe operations and maintenance

- any negative transfer of training from the old removed HSIs to the corresponding new ones was identified and corrected

- no new problems are created by coordinating tasks between the remaining old HSIs and new HSIs

- no unanticipated negative effects on personnel interaction and teamwork have surfaced

12.4.2.2 Modernization Programs Consisting of Many Small Modifications

(1)   The applicant should assure that each modification follows an HFE program that provides standardization and consistency (1) between old and new equipment, and (2) across the new systems being implemented.

(2)   The applicant should verify that new modifications fulfill a clear operational need, and do not interfere with existing systems.

Additional Information: For example, the auditory alerts in a new HSI should not distract operators from addressing more important alarms.

12.4.2.3 Modernization Programs Consisting of Large Modifications during Multiple Outages

(1)   Interim configurations may exist for long times (e. g., a refueling cycle), and therefore, applicants should verify that they are acceptable from both engineering and operations perspectives and that they meet regulatory requirements. The applicant's evaluations should include:

- PRA evaluations to ensure minimizing high-risk situations

- FSAR evaluations to assure defense against design basis accidents

- technical-specifications evaluations to determine if changes are needed

- defense in depth evaluations to ensure meeting the criteria in RG 1.174

(2)   The applicant should perform task analysis for each interim configuration to verify that any task demands are known and do not degrade personnel performance.

(3)   The applicant should update the HRA to address any unique tasks that may impact risk, as well as any changes to existing tasks due to the interim configuration.

(4)   The applicant should verify that the HSIs needed to perform important tasks (as defined in Section 6) are consistent and standardized. Personnel should not have to use both old and new HSIs for different aspects of the same task.

(5)   The applicant should develop procedures for temporary configurations of systems and HSIs that personnel use when the plant is not shutdown.

(6)   The applicant should develop training for temporary configurations of systems, HSIs, and procedures that personnel can use when the plant is not shutdown.

(7)   The applicant should consider the following aspects of V&V:

- *HFE Design Verification* – Temporary configurations of the systems, HSIs, and procedures that operations and maintenance personnel employ when the plant is not shutdown should be reviewed to verify that their design is consistent with the principles of good HFE design (e.g., conforms to a plant-specific style guide or NUREG-0700).

- *HSI Task-Support Verification* – Temporary configurations of the systems, HSIs, and procedures, which operations and maintenance personnel may use when the plant is not shutdown, should be reviewed to verify that their design supports the intended tasks.

  Additional Information: For example, if a temporary configuration of plant systems introduces special monitoring requirements, then the HSIs should give the necessary information.

- *ISV* - Interim configurations should be validated if so warranted by the risk-significance of the personnel tasks affected by them.

12.4.2.4 Modernization Programs Where both Old and New Equipment are Left in Place

(1)   The applicant should identify and address negative effects on personnel performance due to control room or HSI clutter resulting from using old and new HSIs in parallel.

(2)   The applicant should identify and address negative effects on personnel performance resulting from the simultaneous presence of parallel alarms.

(3)   The applicant should identify and address negative effects on personnel performance resulting from differences in information from old and new systems on the same parameter or equipment.

(4)   The applicant should identify and address any safety concerns from providing controls that operators can access from two different HSIs.

      Additional Information: For example, a switch may be installed to select which HSI will control the equipment, thus preventing simultaneous control inputs.

12.4.2.5 Modernization Programs Where New Non-functional HSIs are in Place in Parallel with Old Functional HSIs

(1)   The applicant should evaluate the potential for negative effects on personnel performance due to control room or HSI clutter resulting from having old and new HSIs available in parallel. Where safety concerns are identified, the applicant should take measures to improve the HSIs.

(2)   The applicant should ensure that the non-functional state of HSIs is clearly indicated.

12.5   Bibliography

10 CFR Part 50.59: *U.S. Code of Federal Regulations*, Part 50.59 *"Changes, Tests, and Experiments*," Title 10, "Energy."

IEC 62096 *Nuclear Power Plants - Instrumentation and Control: Guidance for the Decision on Modernization* (IEC, 2009).

NEI 96-01, Rev. 1: *Guidelines for 10 CFR 50.59 Implementation* (Nuclear Energy Institute, 2000).

Regulatory Guide 1.174, Rev. 2: *An Approach for using Probabilistic Risk Assessment in Risk-Informed Decisions on Plant-Specific Changes to the Licensing Basis* (NRC, 2011).

13   HUMAN PERFORMANCE MONITORING

13.1   Background

A human performance monitoring program will help to provide reasonable assurance that the confidence developed by completing a thorough HFE program, culminating in a verification and validation of the control room and integrated systems design, is maintained over time. The NRC staff does not intend that licensees periodically repeat the validation of the fully integrated system; however, there should be sufficient evidence to be reasonably confident that personnel have maintained the skills needed to complete the actions that their training requires. Additionally, RG 1.174, Section C.3, Element 3, related to plant modifications, discusses an Implementation and Monitoring Program, to ensure that no unexpected safety degradation occurs due to changes to the plant's licensing basis, made via this RG. This aspect of monitoring is covered below in Section 13.4, Criterion 2.

13.2   Objective

The NRC staff uses the review criteria in this section to verify that the applicant prepared a human performance monitoring program to:

- adequately assure that the conclusions drawn from the ISV remain valid with time
- ensure that no significant safety degradation occurs because of any changes made in the plant

The applicant may incorporate this monitoring program into their problem identification and resolution program and their training program.

13.3   Applicant Products and Submittals

The product of the applicant's human performance monitoring element is a monitoring program to use throughout the life of the facility.

The applicant should provide for review an IP for monitoring human performance after the plant becomes operational. Submittal of an RSR is not expected because a problem identification and resolution program will be established as part of normal plant operations and so will be subject to routine NRC inspections.

13.4   Review Criteria

(1)   The scope of the applicant's performance monitoring program should provide reasonable assurance that:

- personnel can use the design effectively, including within the control room and between the control room, local control stations, and support centers
- changes made to the HSIs, procedures, and training do not adversely affect human performance, e.g., they do not interfere with previously trained skills
- important human actions can be accomplished within the criteria for time and performance
- an acceptable level of performance, established during ISV, is maintained

103

(2)   The applicant should develop and document a human performance monitoring program. The program should:

- be able to trend human performance after the plant is operational, or after modifications were made to demonstrate that performance is consistent with that assumed in the various analyses that were conducted to justify the change

- begin at initial loading of the plant's fuel

    Additional Information: Applicants may integrate, or coordinate, their performance monitoring for risk-informed changes, made using RG 1.174, with existing programs for monitoring personnel performance, such as the program for licensed operator training, and the corrective action program. Also, if a plant change requires monitoring of actions that were not included in existing training programs, it may be advantageous to adjust the existing program rather than to develop additional monitoring programs for risk-informed purposes.

(3)   The applicant should structure the program such that:

- the level of monitoring human actions is commensurate with their safety importance

- feedback of information and corrective actions are accomplished in a timely manner

- degradations in performance can be detected and corrected before they compromise plant safety (e.g., by use of the plant's simulator during periodic training exercises)

(4)   The performance of the plant or personnel under actual design basis conditions may not be readily measurable. When these conditions cannot be simulated, monitored, or measured, the applicant should use available information that most closely approximates performance data under actual conditions.

(5)   The applicant should include in the program provisions for determining the specific cause of performance degradation and failures, undertaking corrective actions, and trending them. Specifically, the program should:

- define and address the significance of failure, the circumstances surrounding failure or degraded performance, characteristics of the failure, and whether the failure is isolated or has generic or common-cause implications

- for significant failures and degradations, the program should identify the cause and stipulate the corrective actions necessary to preclude repetitions

- identify and ensure the implementation of any corrective actions necessary to preclude the recurrence of unacceptable failures or degraded performance.

- contain provisions for trending performance degradation and failures

13.5 Bibliography

CN Number 02-001: *NRC Inspection Manual: Chapter 2515, Light-water Reactor Inspection Program - Operations Phase* (NRC, 2002).

CN Number 01-015: *NRC Inspection Manual: Chapter 0609, Significance Determination Process* (NRC, 2001).

IP 71715: *Sustained Control Room and Plant Observation.* (NRC, periodically updated).

NUREG-1649: *Reactor Oversight Process* (NRC, 2000).

Regulatory Guide 1.174: *An Approach for Using Probabilistic Risk Assessment in Risk-Informed Decisions on Plant-Specific Changes to the Licensing Basis* (NRC, 2011).

14   REFERENCES

American Institute of Aeronautics and Astronautics (AIAA) (2001). *Guide to Human Performance Measurements* (ANSI/AIAA G-035A-2000). Washington, DC: American National Standards Institute.

American Nuclear Society (1999). *Selection, Qualification, and Training of Personnel for Nuclear Power Plants* (ANSI/ANS 3.1-1993, R1999). Washington, DC: American National Standards Institute.

American Nuclear Society (1994). *Time Response Design Criteria for Safety-Related Operator Actions* (ANSI/ANS 58.8-1994). Washington, DC: American National Standards Institute.

American Nuclear Society (2006). *Administrative Controls and Quality Assurance for the Operational Phase of NPPs* (ANS 3.2-2006). LaGrange Park, IL: American Nuclear Society.

American Nuclear Society (2009). *Nuclear Power Plant Simulators for Use in Operator Training* (ANSI/ANS-3.5-2009). Washington, DC: American National Standards Institute.

Barnes, V., Moore, C., Wieringa, D., Isakson, C., Kono, B. & Gruel, R. (1989). *Techniques for Preparing Flowchart Format Emergency Operating Procedures* (NUREG/CR-5228, Volumes 1 and 2). Washington, DC: U.S. Nuclear Regulatory Commission.

Bongarra, J., Van Cott, H., Pain, R., Peterson, L. & Wallace, R. (1985). *Human Factors Design Guidelines for Maintainability of Department of Energy Nuclear Facilities* (Tech. Report UCRL-15673). Livermore, CA: Lawrence Livermore National Laboratory.

Brown, W., O'Hara, J. & Higgins, J. (2000). *Advance Alarm Systems: Guidance Development and Technical Basis* (NUREG/CR-6684). Washington, DC: U.S. Nuclear Regulatory Commission.

Burgy, D., Lempges, C., Miller, A., Schroeder, L. Van Cott, H. & Paramore, B. (1983). *Task Analysis of Nuclear Power Plant Control Room Crews* (NUREG/CR-3371, Volumes 1 and 2). Washington, DC: U.S. Nuclear Regulatory Commission.

Gertman, D., Hallbert, B., Parrish, W., Sattision, M., Brownson, D. & Tortorelli, J. (2001). *Review of Findings for Human Performance Contribution to Risk in Operating Events* (NUREG/CR-6753). Washington, DC: U.S. Nuclear Regulatory Commission.

Higgins, J. & Nasta, K. (1996). *HFE Insights For Advanced Reactors Based Upon Operating Experience* (NUREG/CR-6400). Washington, DC: U.S. Nuclear Regulatory Commission.

Higgins, J. & O'Hara, J. (2000). *Proposed Approach for Reviewing Changes to Risk-Important Human Actions* (NUREG/CR-6689). Washington, DC: U.S. Nuclear Regulatory Commission.

Higgins, J., O'Hara, J., Lewis, P., Persensky, J., Bongarra, J., Cooper, S. & Parry, G. (2007). *Guidance for the Review of Changes to Human Actions* (NUREG-1764, Rev. 1). Washington, DC: U.S. Nuclear Regulatory Commission.

Human Factors and Ergonomic Society (HFES) (2007). *Human Factors Engineering of Computer Workstations* (ANSI/HFES 100-2007). Washington, DC: American National Standards Institute.

Human Factors and Ergonomic Society (HFES) (2008). *Human Factors Engineering of Software User Interfaces* (ANSI/HFES 200). Washington, DC: American National Standards Institute.

Institute of Electrical and Electronics Engineers (2004). *IEEE Recommended Practice for the Application of Human Factors Engineering to Systems, Equipment, and Facilities of Nuclear Power Generating Stations and Other Nuclear Facilities* (IEEE Std. 1023-2004). New York: Institute of Electrical and Electronics Engineers.

Institute of Electrical and Electronics Engineers (1999). *IEEE Guide to the Evaluation of Human-System Performance in Nuclear Power Generating Stations* (IEEE Std. 845-1999). New York: Institute of Electrical and Electronics Engineers.

Institute of Electrical and Electronics Engineers (1997). *A Guide for Incorporating Human Action Reliability Analysis for Nuclear Power Generating Stations* (IEEE Std. 1082-1997). New York: Institute of Electrical and Electronics Engineers.

International Atomic Energy Agency (1999). *Basic Safety Principles for Nuclear Power Plants* (Safety Series No. 75-INSAG-3). Vienna, Austria: International Atomic Energy Agency

International Atomic Energy Agency (1992). *The Role of Automation and Humans in Nuclear Power Plants* (IAEA-TECDOC-668). Vienna, Austria: International Atomic Energy Agency.

International Electrotechnical Commission (2009). *Nuclear Power Plants - Control Rooms - Design* (IEC 60964, Edition 2.0). Geneva, Switzerland: Bureau Central de la Commission Electrotechnique Internationale.

International Electrotechnical Commission (2009). *Nuclear Power Plants - Instrumentation and Control: Guidance for the Decision on Modernization* (IEC 62096). Geneva, Switzerland: Bureau Central de la Commission Electrotechnique Internationale.

International Electrotechnical Commission (2000). *Nuclear Power Plants - Design of Control Rooms - Functional Analysis and Assignment* (Standard No. 61839). Geneva, Switzerland: Bureau Central de la Commission Electrotechnique Internationale.

International Electrotechnical Commission (1995). *Nuclear Power Plants Main Control Rooms-- Verification and Validation of Design* (IEC Standard 1771). Geneva, Switzerland: Bureau Central de la Commission Electrotechnique Internationale.

International Standards Organization (2006). *Ergonomic Design of Control Centres: Part 7: Principles for the Evaluation of Control Centres* (ISO 11064-7). Geneva, Switzerland: International Standards Organization.

International Standards Organization (2000). *Ergonomic Design of Control Centres - Part 1: Principles for the Design of Control Centres* (ISO 11064-1). Geneva, Switzerland: International Standards Organization.

Kirwan, B. & Ainsworth L. (Eds) (1992). *A Guide to Task Analysis*. London: Taylor and Francis.

Kolaczkowski, A., Forester, J., Gallucci, R., Klein, A., Bongarra, J. Qualls, P. Barbadoro, P. & Lois, E. (2007). *Demonstrating the Feasibility and Reliability of Operator Manual Actions in Response to Fire* (NUREG-1852). Washington, DC: U.S. Nuclear Regulatory Commission.

Nuclear Energy Institute (2005). *10 CFR 50.59 SSC Categorization Guideline* (NEI 00-04). Washington, DC: Nuclear Energy Institute.

Nuclear Energy Institute (2000). *Guidelines for 10 CFR 50.59 Implementation* (NEI 96-01, Rev. 1). Washington, DC: Nuclear Energy Institute.

NRC (periodically updated). *Sustained Control Room and Plant Observation* (IP 71715). Washington, DC: U.S. Nuclear Regulatory Commission.

NRC (periodically updated). *Plant Procedures* (IP 42700). Washington, DC: U.S. Nuclear Regulatory Commission.

NRC (periodically updated). *Emergency Operating Procedures* (IP 42001). Washington, DC: U.S. Nuclear Regulatory Commission.

NRC (periodically updated). *Training and Qualification Effectiveness* (IP 41500). Washington, DC: U.S. Nuclear Regulatory Commission.

NRC (Revised periodically). *Training Review Criteria and Procedures* (NUREG-1220). Washington, DC: U.S. Nuclear Regulatory Commission.

NRC (2011). *An Approach for Plant-Specific, Risk-Informed Decision Making: Technical Specifications* (Regulatory Guide 1.177). Washington, DC: U.S. Nuclear Regulatory Commission.

NRC (2011). *An Approach for Using Probabilistic Risk Assessment in Risk-Informed Decisions on Plant-Specific Changes to the Licensing Basis* (Regulatory Guide 1.174). Washington, DC: U.S. Nuclear Regulatory Commission.

NRC (2011). *Nuclear Power Plant Simulators for Use in Operator Training* (Regulatory Guide 1.149, Rev. 4). Washington, DC: U.S. Nuclear Regulatory Commission.

NRC (2010). *Bypassed and Inoperable Status Indication for Nuclear Power Plant Safety Systems* (Regulatory Guide 1.47, Rev 1). Washington, D.C.: U. S. Nuclear Regulatory Commission.

NRC (2010). *Manual Initiation of Protective Actions* (Regulatory Guide 1.62). Washington, DC: U.S. Nuclear Regulatory Commission.

NRC (2009). *Digital Instrumentation and Controls, Task Working Group 2: Diversity and Defense-in-Depth Issues, Interim Staff Guidance* (DI&C-ISG-02, Rev. 2). Washington, DC: U.S. Nuclear Regulatory Commission.

NRC (2009). *Communications between the NRC and Reactor Licensees during Emergencies and Significant Events* (RIS 2009-10). Washington, DC: U.S. Nuclear Regulatory Commission.

NRC (2008). *Highly-Integrated Control Rooms-Human Factors Issues* (DI&C-ISG-05 Revision 1). Washington, D.C.: U. S. Nuclear Regulatory Commission.

NRC (2007). *Standard Review Plan, Chapter 18 - Human Factors Engineering* (NUREG-0800). Washington, DC: U.S. Nuclear Regulatory Commission.

NRC (2007). *Guidance for Evaluation of Diversity and Defense-In-Depth in Digital Computer-Based Instrumentation and Control Systems* (NUREG-0800, Branch Technical Position 7-19, Rev. 5). Washington, D.C.: U. S. Nuclear Regulatory Commission.

NRC (2007). *Knowledge & Abilities Catalogue for NPP Operators: BWRs* (NUREG-1123). Washington, DC: U.S. Nuclear Regulatory Commission.

NRC (2007). *Meteorological Measurements Programs in Support of Nuclear Power Plants* (Regulatory Guide 1.23, Rev 1). Washington, D.C.: U. S. Nuclear Regulatory Commission.

NRC (2007). *Operator Licensing Examination Standards for Power Reactors* (NUREG-1021, Rev. 9, Supplement 1). Washington, DC: U.S. Nuclear Regulatory Commission.

NRC (2006). *Criteria for Accident Monitoring Instrumentation for Nuclear Power Plants* (Regulatory Guide 1.97, Rev 4). Washington, D.C.: U. S. Nuclear Regulatory Commission.

NRC (2006). *Guidelines for Categorizing Structures, Systems, and Components in Nuclear Power Plants According to Their Safety Significance* (Regulatory Guide 1.201, Rev. 1). Washington, DC: U.S. Nuclear Regulatory Commission.

NRC (2006). *Guidelines for Categorizing Structures, Systems, and Components in Nuclear Power Plants According to Their Safety Significance* (Regulatory Guide 1.201, Rev. 1). Washington, DC: U.S. Nuclear Regulatory Commission.

NRC (2005). *Good Practices for Implementing Human Reliability Analysis (HRA)* (NUREG-1792). Washington, DC: U.S. Nuclear Regulatory Commission.

NRC (2002). *Human System Interface Design Review Guidelines* (NUREG-0700). Washington, DC: U.S. Nuclear Regulatory Commission.

NRC (2002). *NRC Inspection Manual: Chapter 2515, Light-Water Reactor Inspection Program - Operations Phase* (CN Number 02-001). Washington, DC: U.S. Nuclear Regulatory Commission.

NRC (2002). *Standard Review Plan: Chapter 19, Use of Probabilistic Risk Assessment in Plant-Specific, Risk-Informed Decision Making: General Guidance* (NUREG-0800). Washington, D.C.: U.S. Nuclear Regulatory Commission.

NRC (2001). *NRC Inspection Manual: Chapter 0609, Significance Determination Process* (CN Number 01-015). Washington, DC: U.S. Nuclear Regulatory Commission.

NRC (2000). *Reactor Oversight Process* (NUREG-1649). Washington, DC: U.S. Nuclear Regulatory Commission.

NRC (2000). *Personnel Selection and Training* (Regulatory Guide 1.8, Rev 3). Washington, DC: U.S. Nuclear Regulatory Commission.

NRC (2000). *Technical Basis and Implementation Guidelines for a Technique for Human Event Analysis (ATHEANA)* (NUREG-1624, Rev. 1). Washington, DC: U.S. Nuclear Regulatory Commission.

NRC (1999). *Standard Review Plan: Chapter 13, Conduct of Operations* (NUREG-0800). Washington, D.C.: U.S. Nuclear Regulatory Commission.

NRC (1998). *Knowledge & Abilities Catalogue for NPP Operators: PWRs* (NUREG-1122). Washington, DC: U.S. Nuclear Regulatory Commission.

NRC (1997). *Crediting of Operator Actions in Place of Automatic Actions and Modifications of Operator Actions, Including Response Times* (Information Notice 97-78). Washington, DC: U.S. Nuclear Regulatory Commission.

NRC (1997). *Individual Plant Examination Program: Perspectives on Reactor Safety and Plant Performance* (NUREG-1560). Washington, DC: U.S. Nuclear Regulatory Commission.

NRC (1995). *Results of Shift Staffing Study* (Information Notice 95-48). Washington, DC: U.S. Nuclear Regulatory Commission.

NRC (1992). *Lessons Learned From the Special Inspection Program for Emergency Operating Procedures* (NUREG-1358, Supplement 1). Washington, DC: U.S. Nuclear Regulatory Commission.

NRC (1989). *A Status Report Regarding Industry Implementation of Safety Parameter Display System* (NUREG-1342). Washington, D.C.: U. S. Nuclear Regulatory Commission.

NRC (1989). *Lessons Learned from the Special Inspection Program for Emergency Operating Procedures* (NUREG-1358). Washington, DC: U.S. Nuclear Regulatory Commission.

NRC (2008). *Guidance to Operators at the Controls and to Senior Operators in the Control Room of a Nuclear Power Unit* (Regulatory Guide 1.114, Rev. 2). Washington, DC: U.S. Nuclear Regulatory Commission.

NRC (1982). *Guidelines for the Preparation of Emergency Operating Procedures* (NUREG-0899). Washington, DC: U.S. Nuclear Regulatory Commission.

NRC (1982). *Policy on Factors Causing Fatigue of Operating Personnel at Nuclear Reactors* (Generic Letter No. 82-12). Washington, DC: U.S. Nuclear Regulatory Commission.

NRC (1981). *Human Factors Acceptance Criteria for the Safety Parameter Display System* (NUREG-0835). Washington, D.C.: U. S. Nuclear Regulatory Commission

NRC (1981). *Functional Criteria for Emergency Response Facilities* (NUREG-0696). Washington, D.C.: U. S. Nuclear Regulatory Commission.

NRC (1980). *Criteria for Preparation and Evaluation of Radiological Emergency Response Plans and Preparedness in Support of Nuclear Power Plants* (NUREG-0654/FEMA-REP-1, Rev. 1). Washington, D.C.: U. S. Nuclear Regulatory Commission.

NRC (1980). *Clarification of TMI Action Plan Requirements* (NUREG-0737 and supplements). Washington, DC: U.S. Nuclear Regulatory Commission.

NRC (1978). *Quality Assurance Program Requirements* (Regulatory Guide 1.33, Revision 2). Washington, DC: U.S. Nuclear Regulatory Commission.

O'Hara, J. & Higgins, J. (2010). Human-System Interfaces to Automatic Systems: Review Guidance and Technical Basis (BNL Technical Report 91017-2010). Upton, NY: Brookhaven National Laboratory.

O'Hara, J., Higgins, J. & Kramer, J. (2000). *Advanced Information Systems: Technical Basis and Human Factors Review Guidance* (NUREG/CR-6633). Washington, DC: U.S. Nuclear Regulatory Commission.

O'Hara, J., Higgins, J., Stubler, W. & Kramer, J. (2000). *Computer-Based Procedure Systems: Technical Basis and Human Factors Review Guidance* (NUREG/CR-6634). Washington, DC: U.S. Nuclear Regulatory Commission.

O'Hara, J., Stubler, W., Brown, W. & Higgins, J. (1997). *Integrated System Validation: Methodology and Review Criteria* (NUREG/CR-6393). Washington, DC: U.S. Nuclear Regulatory Commission.

Persensky, J., Szabo, A., Plott, C., Engh, T. & Barnes, A. (2005). Guidance for Assessing Exemption Requests from the Nuclear Power Plant Licensed Operator Staffing Requirements Specified in 10 CFR *50.54(m)* (NUREG-1791). Washington, D.C.: U.S. Nuclear Regulatory Commission.

Plott, C., Engh, T. & Barnes, V. (2004). *Technical Basis for Regulatory Guidance for Assessing Exemption Requests from Nuclear Power Plant Licensed Operator Staffing Requirements Specified in 10 CFR 50.54(m)* (NUREG/CR-6838). Washington, D.C.: U.S. Nuclear Regulatory Commission.

Price, H., Maisano, R. & Van Cott, H. (1982). *The Allocation of Functions in Man-Machine Systems: A Perspective and Literature Review* (NUREG/CR-2623). Washington, DC: U.S. Nuclear Regulatory Commission.

Pulliam, R., Price, H., Bongarra, J., Sawyer, C. & Kisner, R. (1983). *A Methodology for Allocation of Nuclear Power Plant Control Functions to Human and Automated Control* (NUREG/CR-3331). Washington, DC: U.S. Nuclear Regulatory Commission.

Shraagen, J., Chipman, S. & Shalin, V. (2000). *Cognitive Task Analysis*. New Jersey: Lawrence Erlbaum Associates.

Stubler, W., O'Hara, J., Higgins, J. & Kramer, J. (2000). *Human-System Interface and Plant Modernization Process: Technical Basis and Human Factors Review Guidance* (NUREG/CR-6637). Washington, DC: U.S. Nuclear Regulatory Commission.

Stubler, W. & O'Hara, J. (1996). *Group-View Displays: Functional Characteristics and Review Criteria* (BNL TR E2090-T4-4-12/94, Rev. 1). Upton, New York: Brookhaven National Laboratory.

Stubler, W., O'Hara, J. & Kramer, J. (2000). *Soft Controls: Technical Basis and Human Factors Review Guidance* (NUREG/CR-6635). Washington, DC: U.S. Nuclear Regulatory Commission.

Stubler, W., Higgins, J. & Kramer, J. (2000). *Maintenance of Digital Systems: Technical Basis and Human Factors Review Guidance* (NUREG/CR-6636). Washington, DC: U.S. Nuclear Regulatory Commission.

Stubler, W. & O'Hara, J., (1996). *Human-System Interface Design Process and Review Criteria* (BNL TR E2090-T4-1-9/96). Upton, New York: Brookhaven National Laboratory.

U.S. Code of Federal Regulations (revised periodically), Part 2.802, "Petition for Rulemaking" Title 10, "Energy," Washington, DC: U.S. Government Printing Office.

U.S. Code of Federal Regulations (revised periodically), Part 20, "Standards for Protection Against Radiation" Title 10, "Energy," Washington, DC: U.S. Government Printing Office.

U.S. Code of Federal Regulations (revised periodically), Part 26, "Fitness for Duty Programs," Title 10, "Energy," Washington, DC: U.S. Government Printing Office.

U.S. Code of Federal Regulations (revised periodically), Part 50, "Domestic Licensing of Production and Utilization Facilities," Title 10, "Energy," Washington, DC: U.S. Government Printing Office.

U.S. Code of Federal Regulations (revised periodically), Part 52, "Early Site Permits; Standard Design Certifications; and Combined Licenses for Nuclear Power Plants," Title 10, "Energy," Washington, DC: U.S. Government Printing Office.

U.S. Code of Federal Regulations (revised periodically), Part 55, "Operator's Licenses," Title 10, "Energy," Washington, DC: U.S. Government Printing Office.

Vicente, K., (1999). *Cognitive Work Analysis: Toward Safe, Productive, and Healthy Computer-Based Work*. New Jersey: Lawrence Erlbaum Associates.

Wieringa, D., Moore, C. & Barnes, V. (1998). *Procedure Writing: Principles and Practices*. Columbus, Ohio: Battelle Press.

GLOSSARY

Benchmark-referenced performance criteria - Performance is compared with criteria established using a benchmark system, e.g., a current system predefined as acceptable. (See also criteria for requirement-referenced, normative-referenced, and expert-judgment-referenced performance criteria).

Bias - Bias is an aspect of an evaluation methodology that systematically modifies performance or its interpretation.

Component - The meaning of the word component depends on its context. In that of the entire plant, it is an individual piece of equipment, such as a pump, valve, or vessel; usually part of a plant system. In the context of an HSI, a component is one part of a larger unit, such as one meter on a control board. In a maintenance context, a component is a subdivision of a unit of equipment that the maintainer can treat as an object, but which can be further broken down into parts; for example, a mounting board together with its mounted parts is a component.

Concept of operations - A concept of operations (ConOps) defines the goals and expectations for the new system from the perspective of users and other stakeholders and defines the high-level considerations to address as the detailed design evolves. An HFE-focused ConOps addresses the following six dimensions:

- Plant Goals (or Missions)
- Agents' Roles and Responsibilities
- Staffing, Qualifications, and Training
- Management of Normal Operations
- Management of Off-normal Conditions and Emergencies
- Management of Maintenance and Modifications

Concept of use - A concept of use describes how human system interface (HSI) is used to support plant operations; i.e., it describes the capabilities and functions of the HSIs and how they support user tasks

Construct validity - The extent to which a selected performance measure accurately represents the aspect of performance to be measured.

Expert-judgment-referenced performance criteria - Performance is compared with criteria established by expert judgment. (See also requirement-referenced, benchmark-referenced, and normative-referenced performance criteria).

Function - (1) A software-supported capability provided to a user to aid in performing a task. (2) A process or activity required to achieve a desired goal; see "safety function."

Function allocation - The process of assigning responsibility for accomplishing functions to personnel or automation, or to a combination of them.

Functional requirements analysis - Functional requirements analysis identifies functions that must be performed to satisfy plant safety objectives and goals; that is, to prevent or mitigate the consequences of postulated accidents that could damage the plant or cause undue risk to the health and safety of the public; and to fulfill the plant's goal/mission.

Functional requirements specification - A specification identifying the functions and characteristics that the human-system interface and its components must accomplish or satisfy.

Human-centered design goals - The goals of human factors engineering design that address the cognitive and physical support of personnel performance.

Human factors - A body of scientific facts about human characteristics. The term covers all biomedical, psychological, and psycho-social considerations. It includes, but is not limited to, principles and applications in human factors engineering, personnel selection, job design, training, job performance aids, and human performance evaluation (see "Human factors engineering").

Human factors engineering (HFE) - The application of knowledge about human capabilities and limitations to designing the plant, its systems, and equipment. HFE affords reasonable assurance that the design of the plant, systems, equipment, human tasks, and the work environment are compatible with the sensory, perceptual, cognitive, and physical attributes of the personnel who operate, maintain, and support the plant or other facility (see "Human factors").

HFE-significant I&C degradations - The failure modes and degraded conditions of the I&C system that have the potential to impact HSIs used by personnel to perform important human actions.

Human-system interfaces (HSIs) - A human-system interface is that part of the system through which personnel interact to perform their functions and tasks. Major HSIs include alarms, information displays, controls, and procedures. Their use can be influenced directly by factors such as (1) the organization of HSIs into workstations (e.g., consoles and panels); (2) the arrangement of workstations and supporting equipment into facilities, such as a main control room, remote shutdown station, local control station, technical support center, and emergency-operations facility; and (3) the environmental conditions in which the HSIs are used, including temperature, humidity, ventilation, illumination, and noise. The use of HSIs also can be affected indirectly by other aspects of plant design and operation, such as personnel training, shift schedules, work practices, and management/organizational-factors, such as the plant's safety culture.

Important human actions - Important HAs consist of those actions that meet either risk or deterministic criteria.

- **Risk-important human actions** - Actions defined by risk criteria that plant personnel use to assure the plant's safety. There are absolute and relative criteria for defining risk important actions. For absolute ones, a risk-important action is any action whose successful performance is needed to reasonably assure that predefined risk criteria are met. For relative criteria, the risk-important actions are defined as those with the greatest risk compared to all human actions. The identifications can be made quantitatively from risk analyses, and qualitatively from various criteria, such as concerns about task performance based on considering performance-shaping factors.

- **Deterministically-identified important human actions** - Deterministic engineering analyses typically are completed as part of the suite of analyses in the FSAR/DCD in

114

Chapters 7, Instrumentation & Controls, and 15, Transient and Accident Analyses. These deterministic analyses also often credit human actions.

Integrated system validation - Integrated system validation is an evaluation using performance-based tests to determine whether an integrated system design (i.e., hardware, software, and personnel elements) meets performance requirements and supports the plant's safe operation.

Local control station (LCS) - A personnel interface for process control that is not located in the main control room. This includes multifunction panels, single-function LCSs, such as controls (e.g., valves, switches, and breakers), and displays (e.g., meters) that are operated or consulted during normal, abnormal, or emergency operations.

Mockup - A static representation of a human-system interface (see "Simulator" and "Prototype").

Modification - Any type of change or modernization made to HSIs or plant systems that may influence personnel performance.

Normative-referenced performance criteria - Performance is compared with criteria established from the evaluations of many systems (rather than a single benchmark system). The advantage of this approach is that the same measure can be used in evaluating different designs. (See also, requirement-referenced, benchmark-referenced, and expert-judgment-referenced performance criteria).

Operating experience review (OER) - An OER is a review of previous designs similar to the new design to identify, analyze, and address HFE-related problems and issues, so ensuring the avoidance of any negative features associated with predecessor designs in the current design, while retaining positive features.

Performance-based tests - Tests that involve assessing personnel performance, including subjective opinions, to evaluate a design.

Performance-shaping factors (PSFs) - PSFs are factors that influence human reliability via their effects on performance. They include environmental conditions, the design of human-system interfaces, procedures, training, and supervision.

Personnel safety - Personnel safety relates to preventing individual accidents and injuries of the type regulated by the Occupational Safety and Health Administration.

Plant - The operating unit of a nuclear power station, including the nuclear steam-supply system, the turbine, electrical generator, and all associated systems and components. For a multi-unit plant, the term "plant" refers to all systems and processes associated with the unit's ability to produce electrical power, even though other units might share some systems or portions of systems.

Plant safety - Also called "safe operation of the plant." A general term used herein to denote the technical safety objective as articulated by the International Nuclear Safety Advisory Group of the International Atomic Energy Agency (IAEA) in the "Basic Safety Principles for Nuclear Power Plants" (IAEA, 1988): "To prevent with high confidence accidents in nuclear plants; to verify that, for all accidents taken into account in the design of the plant, even those of very low probability, radiological consequences, if any, would be minor; and to provide reasonable assurance that the likelihood of severe accidents with serious radiological consequences is extremely small."

Primary tasks - Those tasks performed by the personnel to supervise the plant (i.e., monitoring, detection, situation assessment, response planning, and response implementation).

Procedures - Written instructions providing guidance to plant personnel for operating and maintaining the plant, and for handling disturbances and emergency conditions.

Product - The activities performed by applicants for each HFE element result in a variety of products. These products may include implementation plans, detailed analysis results, results summary reports, design descriptions, and actual designs, e.g., the control room HSIs. Some of these products are provided to the NRC to support the HFE review process. These products are referred to as submittals.

Prototype - A dynamic representation of a human-system interface that is not linked to a process model or simulator. A model of an interface that includes the functions and capabilities expected in the final system, though not in a finished form. (See "Simulator" and "Mockup").

Requirement - The term "requirements" is used in two different ways in this document: (1) Requirements established as part of the design process; e.g., design requirements, functional requirements, task requirements; and (2) regulatory requirements identified in Title 10 of the *Code of Federal Regulations*. No regulatory requirements are established in this document.

Requirement-referenced performance criteria - Performance is compared with criteria based on quantified performance requirements; i.e., those for system, subsystem, and personnel performance defined through engineering analysis. (See also benchmark-referenced, normative-referenced, and expert-judgment-referenced performance criteria).

Safety - See "Personnel safety," "Plant safety," "Safety evaluation," "Safety function," "Safety issue," and "Safety-related."

Safety evaluation - The NRC's process of reviewing an aspect of a NPP to verify that it meets requirements and that it will perform as needed to afford reasonable assurance of the plant's safety.

Safety function - Safety functions are those functions serving to verify high-level safety objectives, and often are defined in terms of a boundary or entity important to assuring the plant's integrity, and to preventing the release of radioactive materials. A typical safety function is "reactivity control." A high-level objective, such as impeding the release of radioactive material to the environment, is one that designers strive to achieve through the design of the plant, and that plant personnel endeavor to attain by properly operating the plant. The function often is described without reference to specific plant systems and components, or the level of human- and machine-intervention needed to carry out this action. Functions often are accomplished through some combination of lower-level functions, such as "reactor trip." The process of manipulating lower-level functions to satisfy a higher-level function is defined herein as a control function. During function allocation, the control function is assigned to human and machine elements.

Safety issue - An item identified during plant's design, operation, or review that potentially could affect the plant's safe operation.

Safety-related - A term applied to those NPP structures, systems, and components (SSCs) that prevent or mitigate the consequences of postulated accidents that could pose undue risk to the health and safety of the public (see Appendix B to Part 50 of Title 10 of the U.S. Code of Federal Regulations). The design-basis analyses of the safety analysis report are performed upon these SSCs. They also should be part of a full quality assurance program, in accord with Appendix B of that document.

Secondary tasks - Secondary tasks are those personnel must complete when interfacing with the HSI, such as navigation through computer screens to find a needed display and HSI configuration. Complicated secondary tasks often have negative effects on primary task performance (See primary task).

Simulator - A facility that physically represents the human-system interface configuration, and dynamically represents the operating characteristics and responses of the plant in real time. (see "Mockup" and "Prototype").

Situation awareness - Situation awareness is the degree to which personnel's perception of plant parameters and understanding of the plant's condition corresponds to its actual condition at any given time and influences predictions about future states.

Style guide - A document containing guidelines, tailored so they describe the application of HFE guidance to a specific design, such as for a specific plant control room.

Submittal - An applicant's HFE products that are submitted to the NRC as part of the licensing review process. The DCD and FSAR are two important submittals. For HFE reviews, implementation plans and results summary reports are two important types of submittals used in the review of HFE elements. As part of the NRC's review process, submittals are evaluated and the staff may review other HFE products to supplement the safety review.

System - An integrated collection of plant components and control elements that carry out a function alone, or with other plant systems.

Task - A group of activities with a common purpose, often undertaken in close temporal proximity.

Task analysis - Task analysis is the identification of the specific tasks needed to accomplish all personnel functions, and the information, control and task support required to accomplish those tasks.

Testbed - The environment or facility in which human performance is measured. The testbed typically includes a representation of the human-system interface and may include a process model that can be used in testing human and integrated human-system performance.

Top-down review - A review methodology that follows top-down approach. The review starts at the "top" of the design with high-level plant mission goals that then are broken down into functions allocated to human and system resources. Subsequently, further break down defines the tasks to be performed to accomplish function assignments. The human-system interface is designed to best support job task performance. The detailed design is the "bottom" of this top-down design process.

Trade-off evaluations - Comparisons between design options based on considering human performance, as well as aspects of the design.

Validation - The set of activities to ensure that a system can accomplish its intended use, goals, and objectives in the particular operational environment. (See also "Integrated system validation").

Verification - The process by which the design is evaluated to determine whether it (1) provides the information, controls, and task-support needed to accomplish tasks; and (2) conforms to the HFE design guidance.

Vigilance - The degree to which personnel are alert.

Workload - Workload is comprised of the physical, cognitive, and other demands that tasks place on plant personnel. The impact of one or many of these aspects of workload should be considered in the application of performance measures and while comparing alternative design elements.

APPENDIX – COMPOSITION OF THE HFE DESIGN TEAM

The term "HFE design team" is used generically to refer to the personnel who are responsible for the HFE within the scope of this document. There is no intent to prescribe any particular organizational structure for an applicant, nor is it assumed that HFE is the responsibility of a single organization, or that there is necessarily an organizational unit called the HFE design team. Further, the HFE design team may change with time, for example, when HFE responsibility is reassigned from a vendor to a utility.

The education and related professional experience of the HFE design team's personnel should satisfy the minimum qualification specified below for each area of expertise. Qualifying professional experience (e.g., design, development, analysis) for each area should be directly related to those technologies and techniques that will be part of the HFE design and implementation process.

The design team as a whole must satisfy the professional experience qualifications described below. Therefore, satisfaction of the professional experience requirements associated with a particular skill area may be realized through combining the professional experience of two or more members of the HFE design team who each, individually, satisfy other defined credentials of the particular skill area, but who do not possess all of the specified professional experience. It is recognized that one person may possess multiple skills, and that people may have additional responsibilities beyond the HFE design team.

The following lists the areas of expertise for the HFE design team, with an associated listing of their minimum qualifications, and descriptions of their typical contributions to the HFE design and implementation. These descriptions are provided as examples to further detail the potential value of the various areas of expertise to the process of designing and establishing the HFE program; they are not intended to define the total role for each area of expertise.

(1)   **Technical Project Management**
- Minimum qualifications:
    - Bachelor's degree
    - 5 years of experience in nuclear power plant design or operations
    - 3 years of management experience
- Typical contributions:
    - develop and maintain the schedule for the HFE design process
    - provide a central point-of-contact for managing the HFE design and implementation process

(2)   **Systems Engineering**
- Minimum qualifications:
    - Bachelor of Science degree
    - 4 years of cumulative experience in at least three of the following areas of systems engineering; design, development, integration, operation, and test and evaluation
- Typical contributions:
    - provide knowledge of the purpose, operating characteristics, and technical specifications of major plant systems
    - provide input to HFE analyses, especially function and task analyses

- participate in developing procedures and scenarios for task analysis, validation, and other analyses

(3) **Nuclear Engineering**
- Minimum qualifications:
  - Bachelor of Science degree
  - 4 years of nuclear design, development, test, or operations experience
- Typical contributions:
  - provide knowledge of the processes involved in controlling reactivity and generating power
  - supply input to HFE analyses, especially function and task analyses
  - participate in developing scenarios for task analysis, validation, and other analyses

(4) **Instrumentation and Control (I&C) Engineering**
- Minimum qualifications:
  - Bachelor of Science degree
  - 4 years of experience in designing hardware and software aspects of process control systems
  - experience in at least one of the following areas of I&C engineering: design, power plant operations, and test and evaluation
  - familiarity with the theory and practice of software quality assurance and control
- Typical contributions:
  - provide detailed knowledge of the human-system interface (HSI) design, including control and display hardware selection, design, functionality, and installation
  - provide knowledge of information display design, content, and functionality
  - participate in the designing, developing, testing, and evaluating the HSIs
  - participate in developing scenarios for human reliability analysis (HRA), validation, and other analyses involving failures of the HSI data processing systems
  - provide input to software quality assurance programs

(5) **Architect Engineering**
- Minimum qualifications:
  - Bachelor of Science degree:
  - 4 years of experience in design of power plant control rooms
- Typical contributions:
  - provide knowledge of the overall structure of the plant, including performance requirements, design constraints, and design characteristics of the following: containment building, control room, remote shutdown area, and local control stations
  - provide knowledge of the configuration of plant components within the plant
  - provide input to plant analyses, especially function analysis, task analysis, and the development of scenarios for task analysis and validation

(6) **Human Factors Engineering**
- Minimum qualifications:
  - Bachelor's degree in Human Factors Engineering, Engineering Psychology, or related science

- 4 years of cumulative experience related to the human factors aspects of human-computer interfaces. Qualifying experience should include at least the following activities within the context of large-scale, human-machine systems (e.g., process control): design, development, and test and evaluation
- 4 years of cumulative experience related to the human factors aspects of workplace design. Qualifying experience should include at least two of the following activities: design, development, and test and evaluation
- Typical contributions:
  - provide knowledge of human performance capabilities and limitations, applicable human factors design and evaluation practices, and human factors principles, guidelines, and standards
  - develop and perform human factors analyses and participate in resolving identified problems therein

(7) **Plant Operations**
  - Minimum qualifications:
    - has or has held a senior reactor operator license
    - 2 years of experience in relevant nuclear power plant operations
  - Typical contributions:
    - provide knowledge of operational activities, including task characteristics, HSI characteristics, environmental characteristics, and technical requirements related to operational activities
    - provide knowledge of operational activities supporting HSI activities, such as developing HSIs, procedures, and training programs
    - participate in developing scenarios for HRA evaluations, task analyses, HSI tests and evaluations, validation, and other evaluations

(8) **Computer System Engineering**
  - Minimum qualifications:
    - Bachelor's degree in Electrical Engineering or Computer Science, or graduate degree in other engineering discipline (e.g., Mechanical Engineering or Chemical Engineering)
    - 4 years of experience in the design of digital computer systems and real-time systems applications
    - familiarity with the theory and practice of software quality assurance and control
  - Typical contributions:
    - provide knowledge of data processing associated with displays and controls
    - participate in the designing and selecting computer-based equipment, such as controls and displays
    - participate in developing scenarios for HRA, validation, and other analyses involving failures of the HSI data processing systems

(9) **Plant Procedure Development**
  - Minimum qualifications:
    - Bachelor's degree
    - 4 years of experience in developing procedures for nuclear power plants
  - Typical contributions:
    - provide knowledge of operational tasks and procedure formats, especially as presented in emergency procedure guidelines, and operational procedures of current and predecessor plants

- participate in developing scenarios for HRA evaluations, task analyses, HSI tests and evaluations, validation, and other evaluations
- provide input for developing emergency operating procedures, procedure aids, computer-based procedures, and training systems

(10) **Personnel Training**
- Minimum qualifications:
  - Bachelor's degree
  - 4 years of experience in developing personnel training programs for power plants
  - experience in applying the systems approach to training
- Typical contributions:
  - develop content and format of personnel training programs for licensed and non-licensed plant personnel
  - coordinate training issues arising from activities, such as HRA, HSI design, and procedure design with the training program
  - participate in developing scenarios for HRA evaluations, task analyses, HSI tests and evaluations, validation, and other evaluations

(11) **Systems Safety Engineering**
- Minimum qualifications:
  - Bachelor's degree in Science
  - 4 years of experience in system safety engineering
- Typical contributions:
  - identify safety concerns and perform a system safety hazard analysis
  - provide results of system safety hazard analysis to probabilistic risk assessment/HRA and human factors analyses

(12) **Maintainability/Inspectability Engineering**
- Minimum qualifications:
  - Bachelor's degree in Science
  - 4 years of cumulative experience in at least two of the following areas of power plant maintainability and inspectability engineering activity: design, development, integration, and test and evaluation
  - experience in analyzing and resolving plant system and/or equipment-related maintenance problems
- Typical contributions:
  - provide knowledge of maintenance, inspection, and surveillance activities, including task characteristics, HSI characteristics, human performance demands, environmental characteristics, and technical requirements related to the conduct of these activities
  - support the design, development, and evaluation of the control room and other HSIs throughout the plant to provide reasonable assurance that they can be inspected and maintained to the specified level of reliability
  - provide input in the areas of maintainability and inspectability to the development of procedures and training
  - participate in the development of scenarios for HSI evaluations including task analyses, HSI design tests and evaluations, and validation

(13) **Reliability/Availability Engineering**
- Minimum qualifications:
  - Bachelor's degree

- 4 years of cumulative experience in at least two of the following areas of power plant reliability engineering activity: design, development, integration, and test and evaluation
    - knowledge of computer-based, human-interface systems
- Typical contributions:
    - provide knowledge of plant component and system reliability and availability and assessment methodologies to the HSI development activities
    - participate in human reliability analyses
    - participate in the development of scenarios for HSI evaluations, especially validation
    - provide input to the design of HSI equipment to provide reasonable assurance that it meets reliability goals during operation and maintains the specified level of availability

Alternative personal credentials may be accepted as the basis for satisfying these specific minimum qualifications for team membership. Acceptance of such credentials should be evaluated on a case-by-case basis and approved, documented, and retained by the applicant in auditable files. The following factors are examples of alternative credentials that may be considered acceptable:

- Successful completion of all technical portions of an engineering, technology, or related science baccalaureate program may be substituted for the Bachelor's degree. Successful completion will be determined by a transcript or other certification by an accredited institution. For example, completion of 80 semester credit hours may be substituted for the baccalaureate requirement. The courses should be in technical subjects appropriate and relevant to the skill areas of the HFE design team for which the individual will be responsible.

- Related experience may substitute for education at the rate of six semester credit hours for each year of experience, up to a maximum of 60 credit hours.

- Where course work is related to job assignments, post-secondary education may be substituted for experience at the rate of two years of education for one year of experience. Total credit for post-secondary education should not exceed two years of experience credit.

NRC FORM 335
(12-2010)
NRCMD 3.7

U.S. NUCLEAR REGULATORY COMMISSION

BIBLIOGRAPHIC DATA SHEET

(See instructions on the reverse)

1. REPORT NUMBER (Assigned by NRC, Add Vol., Supp., Rev., and Addendum Numbers, if any.)
NUREG-0711, Rev. 3

2. TITLE AND SUBTITLE

Human Factors Engineering Program Review Model

3. DATE REPORT PUBLISHED

MONTH	YEAR
November	2012

4. FIN OR GRANT NUMBER

N6765

5. AUTHOR(S)

John M. O'Hara*, James C. Higgins*, S.A. Fleger, P.A. Pieringer

6. TYPE OF REPORT

Technical

7. PERIOD COVERED (Inclusive Dates)

8. PERFORMING ORGANIZATION - NAME AND ADDRESS (If NRC, provide Division, Office or Region, U. S. Nuclear Regulatory Commission, and mailing address; if contractor, provide name and mailing address.)

*Nuclear Science and Technology Department
Brookhaven National Laboratory
Upton, New York 11973-5000

U.S. Nuclear Regulatory Commission
Washington DC 20555-0001

9. SPONSORING ORGANIZATION - NAME AND ADDRESS (If NRC, type "Same as above", if contractor, provide NRC Division, Office or Region, U. S. Nuclear Regulatory Commission, and mailing address.)

Division of Risk Analysis
Office of Nuclear Regulatory Research
U.S. Nuclear Regulatory Commission
Washington DC 20555-0001

10. SUPPLEMENTARY NOTES

Stephen A. Fleger, NRC Project Manager

11. ABSTRACT (200 words or less)

This document is used by the staff of the Nuclear Regulatory Commission to review the human factors engineering (HFE) programs of applicants for construction permits, operating licenses, standard design certifications, combined operating licenses, and license amendments. The purpose of these reviews is to verify that the applicant's HFE program incorporates accepted HFE practices and guidelines as described within the twelve elements of an HFE program: HFE Program Management, Operating Experience Review, Functional Requirements Analysis and Function Allocation, Task Analysis, Staffing and Qualifications, Treatment of Important Human Actions, Human-System Interface Design, Procedure Development, Training Program Development, Human Factors Verification and Validation, Design Implementation, and Human Performance Monitoring. Each element encompasses five sections: Background, Objective, Applicant Products and Submittals, Review Criteria, and Bibliography.

12. KEY WORDS/DESCRIPTORS (List words or phrases that will assist researchers in locating the report.)

Human factors, human factors engineering, human factors evaluation, human factors review criteria, nuclear safety, safety review, design certification, design review, design process, human-system interface, man-machine interface, verification and validation

13. AVAILABILITY STATEMENT

unlimited

14. SECURITY CLASSIFICATION

(This Page)

unclassified

(This Report)

unclassified

15. NUMBER OF PAGES

16. PRICE

NUREG-0711, Rev. 3

Human Factors Engineering Program Review Model

November 2012